Understanding Photonics & Quantum Optics

A No-Math Authoritative Primer on Quantum Optics, Quantum Entanglement, Quantum Communication, and Photonic Quantum Computing

C. Louis-Charles Phd

Contents

Introduction

The Reader: This Book Is Written For

Picture this: it is early morning, and you reach for your phone before you have even sat up in bed. The screen lights your face with something like a hundred million tiny photons a second. While you check the weather, a fiber-optic cable under the street is shuttling data to your router as pulses of infrared light, each pulse carrying more information in a single blink than an old copper telephone line could carry in a minute. At the supermarket, an hour later, a laser scans the barcode on your cereal box and retrieves the price in a fraction of a millisecond. The satellite navigation on your phone keeps you on the right road by measuring the travel time of radio signals. This calculation relies on the same physics that governs the behavior of light. By nine in the morning, you have already benefited from photonics dozens of times, and the odds are overwhelming that you could not explain any of it if someone asked.

That is not a criticism. It is simply the state of things. Photonics, the science and technology of light, has quietly become the infrastructure layer of modern civilization, the way steel girders and electrical wiring were the invisible backbone of the twentieth century. And yet the educational materials that explain it sit in one of two unfortunate categories. On one side are popular-science articles that tell you light is fast and mysterious and leave it at that. On the other side are university textbooks written for students who already know calculus, differential equations, and tensor algebra, and who are comfortable with Greek letters stacked four rows deep. Between those two extremes, there is almost nothing.

This book was written for the intelligent, curious adult who lives in that gap and is tired of it. You may be a software engineer who keeps reading about the quantum internet and wants to understand what it actually means in physical terms, not just as a buzzword. You may be a biomedical professional who uses photonic instruments every day and wants to know what is genuinely happening inside the device. You may be a technology entrepreneur trying to decide whether photonic quantum computing is a real near-term opportunity or a decade away from practical use. You may be a policy analyst who needs to brief officials on quantum communication security without pretending to be a physicist. Or you may be a lifelong learner, the kind of person who reads carefully and senses, correctly, that photons are the frontier technology of the coming decade and wants to be genuinely fluent rather than merely name-familiar.

Whatever brought you here, this book makes one clear promise: no equations, no derivations, no intimidating walls of mathematical notation. That promise does not come at the cost of accuracy or depth. The real science is here, explained in plain language, through stories, mechanisms, and analogies that connect new ideas to things you already understand. You will finish this book able to explain, with genuine precision, what a photon is, why laser light is so different from sunlight, how a fiber-optic cable carries a continent's worth of data, what quantum entanglement actually means and why it does not allow faster-than-light communication, and how researchers are beginning to build computers made entirely of light. That is a significant body of knowledge. This book is designed to deliver it accessibly and durably.

What This Book Will Teach You

The ten chapters of this book follow a deliberate progression, moving from the most fundamental concepts outward toward the most sophisticated applications. You do not need to have studied physics to follow the arc. Each chapter builds on the one before it, so the ideas accumulate naturally rather than landing as isolated facts.

The first two chapters establish the foundation. Chapter one asks the deceptively simple question: What is light? The answer, as it turns out, is genuinely strange. Light is both a wave and a particle at the same time, and the way it switches between those behaviors depending on how you observe it is one of the most surprising and well-confirmed facts in all of science. You will meet the photon, the individual packet of light energy, and understand why it is the central actor in everything that follows. Chapter two zooms out to the full electromagnetic spectrum: radio waves, microwaves, infrared, visible light, ultraviolet, X-rays, and gamma rays. These are not different kinds of radiation in any deep sense. They are all the same thing, photons, varying only in their energy and the length of their wavelength. Understanding the spectrum gives you the map before the detailed journey begins.

Chapters three, four, and five move into classical photonics, the century-old mastery of light that already shapes everyday life. Chapter three covers how light behaves when it meets a surface or moves from one material to another: reflection, refraction, and diffraction. These are not just textbook curiosities. They are the principles behind eyeglasses, camera lenses, fiber-optic cables, and optical coatings. Chapter four is dedicated to lasers, one of the most consequential inventions of the twentieth century. You will learn why laser light is so different from ordinary light, what stimulated emission means in plain terms, and why surgeons, manufacturers, and research scientists could not do their jobs without it. Chapter five traces the path of light through fiber-optic cables, from the glass strand itself to the undersea cables that carry most of the world's internet traffic. By the end of chapter five, you will understand why photons replaced electrons as the carriers of long-distance information.

The final five chapters cross into quantum photonics, the frontier where light stops being merely useful and starts being astonishing. Chapter six introduces quantum states of light, including the strange phenomenon of squeezed light, in which uncertainty itself can be redistributed to make measurements more precise. Chapter seven is the heart of the book for many readers: quantum entanglement. You will learn what entanglement actually is, why it does not violate relativity, and why it is the foundation for technologies that have no classical counterpart. Chapter eight covers quantum sensing and imaging, the use of quantum light to detect things that no classical instrument can resolve, from gravitational waves to individual molecules. Chapter nine covers quantum communication and cryptography, including the BB84 protocol for quantum key distribution and the no-cloning theorem, which makes quantum cryptography secure in ways that no software encryption can match. Chapter ten closes the arc with photonic quantum computing, the emerging effort to build quantum computers using photons rather than superconducting circuits, and the real-world state of that effort today.

Diagram I.1 - The Ten-Chapter Arc: From Photon To Photonic Quantum Computer

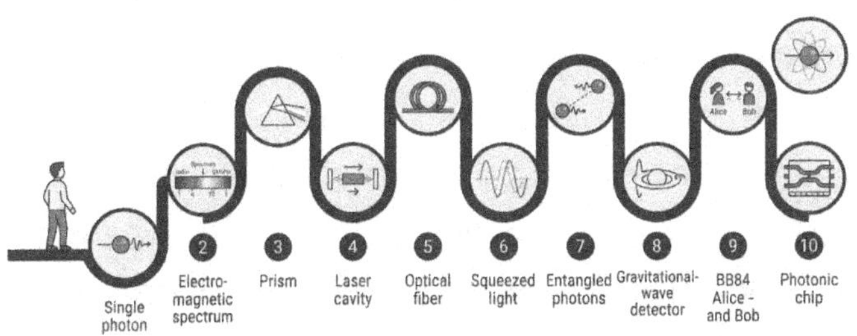

Each chapter adds one layer of understanding, building toward photonic quantum computing.

How To Read This Book

Every chapter opens with a scenario, a short real-world scene set in the present tense that places you inside a moment where the chapter's concept is already at work in the world. A surgeon's prep room before a laser eye procedure. A network operations center monitors terabits of data through a transatlantic fiber cable. A physics laboratory where a researcher fires single photons, one at a time, through a pair of narrow slits. These scenes are not decorative. They are anchors. They give you a concrete human context before the concept is introduced, so that when the idea arrives, it lands somewhere specific and memorable rather than floating in the abstract.

After the opening scenario, each chapter builds its conceptual architecture using three interlocking tools. The first is analogy, connecting a new idea to something you already know. When this book explains why a coherent laser beam and a household light bulb are fundamentally different kinds of light, it will do so by comparing them to a crowd of people walking in different directions versus a marching band moving in perfect step. The second tool is narrative, giving a concept a causal sequence: this leads to that, which produces this, which allows that. Understanding how fiber optics works is not a list of facts to memorize; it is a story about what happens when you trap light inside glass and why the physics of that trapping is so reliable. The third tool is a mechanism that explains how something actually works at the physical level, not just what it is called or what it does. Knowing that a photon has a wavelength is trivia. Understanding why wavelength determines whether a photon passes through glass, bounces off a mirror, or gets absorbed by your skin is genuinely useful knowledge.

You should read the chapters in order on your first pass through the book. The later chapters, particularly chapters six through ten on quantum photonics, assume that you have built up the vocabulary and intuitions from chapters one through five. Jumping to quantum entanglement without understanding what a photon's quantum state is is like reading the final chapter of a novel first: you might follow the words, but you will miss most of the meaning. Once you have read through completely, the chapters work well as standalone references if you want to revisit a specific concept.

A note on terminology: every technical term in this book is defined in plain words the first time it appears. You will encounter real vocabulary, the same terms physicists and engineers use in their work, because vague paraphrases do not give you a real understanding and cannot prepare you for real conversations. But the definitions will always be in ordinary language first, with the formal term provided as a label for something you have already understood. You will never be asked to accept a definition and move on before it makes sense.

There are also more than a hundred original diagrams distributed across the ten chapters. Each diagram is placed at the point in the text where it will do the most work, illustrating a specific concept that is harder to hold in words alone. Pay attention to them. Many of the ideas in photonics are fundamentally spatial and visual, and the diagrams carry genuine information. They are not illustrations of what you have already read; they are a second channel of explanation running in parallel with the text.

A Note On Light

Before you begin the first chapter, it is worth pausing for a moment on what makes light an unusual subject for a primer. Most technology books explain devices or systems that were built by humans and can be redesigned by humans. Light is different. It is a fundamental feature of the universe, not something that engineers invented but something they learned to work with, first by accident and gradually by design.

Light has been studied seriously for more than three centuries. Newton used a prism to show that white light is a mixture of colors. Huygens argued that light was a wave. Maxwell derived the equations showing that light is an electromagnetic wave and predicted its speed from first principles. Einstein won the Nobel Prize for showing that light also behaves as a particle, a photon, capable of knocking electrons out of metal surfaces in ways that a pure wave theory could not explain. Every generation of physicists has found that light contains another layer of depth beneath the layer they had just mastered. That pattern has not stopped.

What is striking about light in the twenty-first century is that it has moved from a subject of pure scientific fascination to the literal infrastructure of civilization. Consider what it means, physically, when you stream a video. A photon of infrared light, generated by a semiconductor laser at a data center, is encoded with digital information and launched into a glass fiber thinner than a human hair. It travels at roughly two-thirds the speed of light in a vacuum, bouncing along the fiber's interior via total internal reflection, covering hundreds of miles in a few milliseconds. It arrives at a detector in your neighborhood, is converted to an electrical signal, and eventually becomes the image on your screen. That is not a metaphor or a simplified summary. That is actually what happens, and it happens billions of times a second for hundreds of millions of simultaneous users.

And yet, fiber optics is the old technology in this story. The frontier uses light not just to carry information but to process it, to perform computation, to generate encryption keys that are secure against any conceivable future computer, and to make measurements precise enough to detect the ripple in spacetime caused by two black holes colliding a billion light-years away. These are not science fiction scenarios. They are things happening in laboratories and, in some cases, in deployed systems right now.

The reason photonics has reached this level of consequence is that light has properties that no other medium can replicate. It travels faster than anything else. It can carry far more information per second than an electrical current through copper wire. It generates almost no heat during transmission. And at the quantum level, individual photons behave in ways that are completely beyond the reach of classical physics, enabling capabilities with no analog in any technology built before the quantum era. Learning how light works is not an optional enrichment activity for the technically curious. It is increasingly a prerequisite for understanding the world being built around you.

This book introduces you to the key vocabulary you will encounter throughout. A photon is the smallest possible unit of light, a discrete packet of electromagnetic energy. Wavelength is the length of one complete wave cycle of light, which determines the color of visible light and the type of radiation for invisible light. Frequency is how many wave cycles pass a fixed point per second: short wavelength means high frequency and higher energy per photon. Coherence describes how well-ordered a beam of light is, whether its photons are synchronized or random. A quantum state is the complete description of a photon's physical condition, including properties like polarization and the direction it is traveling. Entanglement is a relationship between two photons in which measuring one instantly determines a corresponding property of the other, regardless of the distance between them. A qubit is the quantum equivalent of a classical binary bit, and a photon's quantum state can serve as a qubit in a quantum computer. Each of these terms will be explained in much fuller depth in the chapter where it first takes center stage. For now, hold them lightly. They are the cast of characters, and you are about to meet all of them.

Diagram I.2 - How Light Touches Everyday Life

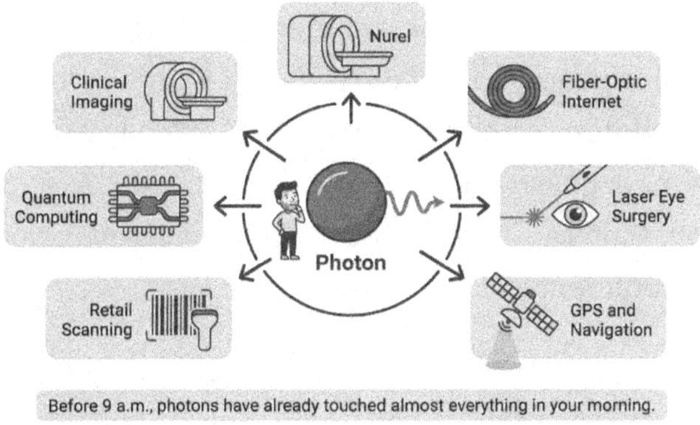

Before 9 a.m., photons have already touched almost everything in your morning.

Takeaway

Light is not a single subject. It is the thread that connects the familiar, a lit screen in a dark bedroom, to the extraordinary, a quantum computer processing information in ways that have no classical equivalent. Understanding light is understanding the physical substrate of the most consequential technologies of this century: fiber-optic communication, precision lasers, quantum cryptography, quantum sensing, and photonic computing.

This book covers that full arc, from the most basic question about what a photon is to the cutting-edge question of how photons might replace transistors as the building blocks of computation. It does so without equations, without an assumed background, and without sacrificing the depth that makes the knowledge genuinely useful.

Every concept you encounter has been worked through so that it arrives as understanding, not just vocabulary. By the time you finish the final chapter, you will have the conceptual fluency to read research journalism about quantum photonics and know what it is actually saying, to ask informed questions in professional contexts, and to evaluate technical claims about photonic technologies with real confidence.

The photon is small. Its effects are not. Let us begin.

1 Light Unveiled: The Photon Story

Opening Scenario

It is past ten o'clock on a weeknight, and you have just remembered something under your bed. You reach for your phone, swipe up, and tap the flashlight icon. A cone of white light floods the floor, you find what you were looking for, and three seconds later, the light is gone. Nothing about that felt like physics.

But here is what was actually happening during those three seconds. The small LED at the back of your phone was emitting light at a rate so staggering that it has no real-world analogy. Each particle traveled from the phone to the floor in less than a billionth of a second. Each carried a precise amount of energy, no more and no less, determined entirely by how fast it was vibrating. None of them had any mass. None of them could be slowed down without being destroyed. Those particles are called photons, and they are the subject of this entire book.

Now imagine sitting near a window on a sunny afternoon and watching the long shafts of light that form when dust drifts through a sunbeam. The light seems to pour in like a slow river, filling the room with warmth and color. It feels continuous and fluid, the way water fills a glass. That feeling is not wrong. Light really does behave like a wave in many circumstances, spreading out, bending around corners, and forming rippling patterns you can see in an oil slick or a soap bubble. Both descriptions, particle and wave, are true at the same time. That apparent contradiction is the starting point for everything quantum optics has to teach us.

Why It Matters

The question of what light actually is might sound like the kind of puzzle that only matters inside a university physics department. It is not. Every technology in this book, from the fiber-optic cables carrying your internet traffic to the photonic quantum computers beginning to emerge from research labs, rests on a precise understanding of photons and how they behave. If you want to understand why a quantum key distribution system is unbreakable in principle, you need to understand that a photon cannot be copied. If you want to understand how a photonic quantum computer works, you need to understand that a photon can exist in a superposition of two states at the same time. The photon is not background material. It is the load-bearing foundation of the entire field.

This chapter traces the intellectual journey that led scientists from the first serious debates about the nature of light to the modern quantum picture. Along the way, we will encounter the experiments and ideas that drove each upgrade in our understanding: Newton's particles, Huygens's waves, Maxwell's electromagnetic revelation, and Einstein's decisive quantum leap. By the time you finish this chapter, you will have a clear and durable mental model of the photon that will make every later chapter easier, richer, and more satisfying to read.

Diagram 1.1 - The Photon Family Portrait

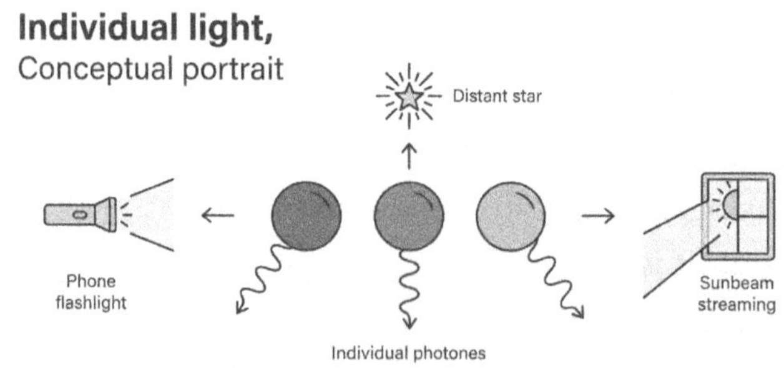

Every source of light, from a phone torch to a distant star, emits its energy as individual photons.

The Particle Versus Wave Debate

For most of human history, nobody had a strong theory about what light actually was. People knew it traveled in straight lines, that it could be reflected by mirrors and bent by glass, and that it came in different colors. Its deeper nature remained wide open. The first serious scientific attempt to answer the question came from two brilliant minds working at roughly the same time, arriving at completely different conclusions.

Isaac Newton, working in the second half of the seventeenth century, proposed that light was made of tiny particles he called corpuscles. Light casts sharp shadows, suggesting it travels in straight lines, as a stream of small balls would. It reflects off mirrors cleanly, just as a ball bounces off a wall. When light passes from air into denser glass or water, it bends toward the surface normal, an observation that Newton explained by suggesting that denser materials exert an attractive pull on the corpuscles, speeding them up and bending their paths. His particle theory was tidy, mechanical, and consistent with the dominant physics of the day.

Christiaan Huygens, a Dutch scientist working in the same era, saw things very differently. He noticed that light behaves in ways that particles cannot explain. When two beams of light cross each other, they pass straight through without deflecting, exactly what waves do. Light bends around the edges of obstacles and spreads out after passing through narrow gaps, a phenomenon known as diffraction. Huygens developed a wave theory in which every point on a wavefront acts as the source of smaller secondary waves that reinforce or cancel each other to produce the observed patterns.

The decisive experimental blow to Newton's theory came in 1801, when the English physicist Thomas Young shone a narrow beam of light through two closely spaced slits in an opaque screen. Behind the screen, he observed not two bright patches but a series of alternating bright and dark bands: an interference pattern. Interference is a signature of waves, not particles. Two waves reinforce each other where their peaks coincide, producing a bright band, and cancel each other where a peak meets a trough, producing a dark band. Two streams of particles could not possibly produce this effect. Young's experiment seemed to settle the debate in favor of waves.

Diagram 1.2 - Newton Versus Huygens: Two Views of Light

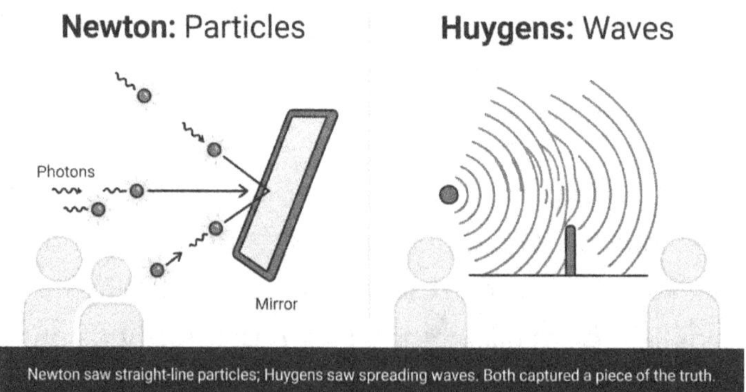

The wave camp received reinforcement from Augustin-Jean Fresnel, who worked out the mathematics of wave interference precisely. A skeptic named Poisson calculated that the theory predicted something absurd: a bright spot of light at the very center of a circular disk's shadow. Dominique Arago ran the experiment and found the bright spot exactly where the theory predicted. The wave theory had foreseen something no one had ever seen, and experiment confirmed it.

What the wave theorists did not yet know was that they had only won half the argument. The particle model was dormant, not dead. It awaited an anomaly the wave theory could not explain, one that would surface in observations about what happens when light shines on a metal surface.

Maxwell's Electromagnetic Revelation

By the middle of the nineteenth century, electricity and magnetism seemed like separate phenomena. Electricity governed lightning and charged objects; magnetism governed compass needles and iron filings. But a growing body of experiments hinted at a connection. Hans Christian Oersted discovered that an electric current running through a wire could deflect a compass needle. Michael Faraday showed that a moving magnet could induce an electric current in a nearby coil. The two forces were influencing each other, and the connection was not coincidental.

James Clerk Maxwell, a Scottish mathematical physicist working in the 1860s, took these scattered threads and wove them into one of the most consequential theories in the history of science. He wrote down a set of equations describing how electric and magnetic fields behave and found that a changing electric field generates a magnetic field. A changing magnetic field generates an electric field. Each field sustains the other in a self-reinforcing cycle that can propagate through space as a wave. The two fields oscillate at right angles to each other and to the direction of travel.

Maxwell could calculate the speed of this electromagnetic wave, because both the electric and magnetic properties of space had been measured by other experimenters. When he did the calculation, he got approximately 300,000 kilometers per second. The measured speed of light was also approximately 300,000 kilometers per second. Maxwell drew the only reasonable conclusion: light is an electromagnetic wave. Two of the greatest mysteries of classical physics turned out to be the same mystery in disguise.

Diagram 1.3 - The Electromagnetic Wave: Oscillating Fields in Space

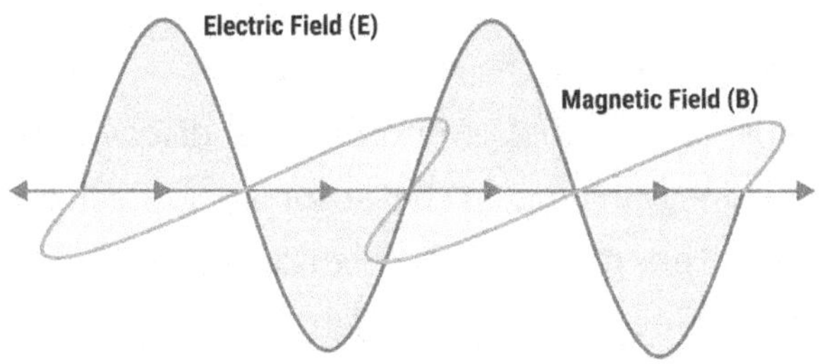

An electromagnetic wave is two fields feeding each other, traveling together at the speed of light.

Maxwell's theory also had a practical consequence. If electromagnetic waves could be generated by oscillating currents, they could be harnessed for communication. The entire modern wireless world, from radio to Wi-Fi to cellular networks, flows directly from Maxwell's equations. When your phone connects to a cell tower, it does so by generating and receiving electromagnetic waves whose behavior Maxwell described more than 150 years ago.

Yet Maxwell's wave theory was heading toward a crisis. Light did not always behave like a continuous wave. In certain experiments, it behaved as though it arrived in concentrated packets rather than spreading out smoothly. The resolution would come from an experiment about what happens when light strikes a metal surface, and it would change physics forever.

Einstein's Photoelectric Effect and the Birth of the Photon

In the 1880s and 1890s, experimenters discovered that when light shines on certain metal surfaces, electrons are ejected from the metal. This phenomenon, called the photoelectric effect, seemed straightforward at first. But the details refused to fit the wave model. If light is a wave, a brighter beam should kick electrons out more energetically than a dimmer beam. That is exactly what happens with most classical wave phenomena: bigger waves mean bigger effects.

The experiments told a completely different story. The energy of the ejected electrons did not depend on the brightness of the light at all. It depended only on the color of the light, which is to say, on the frequency of the wave. Dim blue light ejected electrons with more energy than bright red light. Increasing the brightness of red light merely ejected more electrons, each one with the same low energy as before. Below a certain threshold frequency, no electrons were ejected at all, regardless of how bright the light was. A wave builds up energy gradually, so a sufficiently bright wave at any frequency should eventually give an electron enough energy to escape. But that never happened.

In 1905, Albert Einstein published a paper that explained the anomaly precisely. He proposed that light does not deliver its energy continuously, the way a wave soaks into a surface, but in discrete packets, each carrying an amount of energy proportional to the light's frequency. He borrowed the term "quantum," meaning a discrete unit, from Max Planck, who had introduced the idea a few years earlier. Each quantum delivers its entire packet of energy to one electron at once. If the packet is too small, the electron cannot escape, no matter how many packets arrive per second. If the packet is large enough, the electron escapes with energy to spare. This quantum of light was eventually given its own name: the photon. Einstein's 1905 paper is the work for which he received the Nobel Prize in Physics in 1921.

Diagram 1.4 - The Photoelectric Effect: Packets of Energy Knock Out Electrons

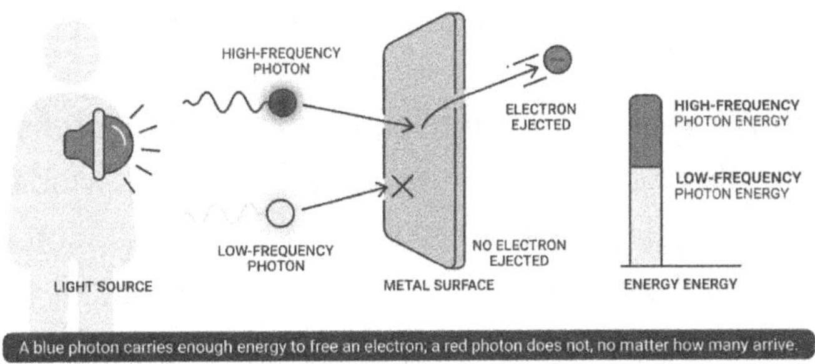

Max Planck proposed his quantum theory in 1900 to explain why heated objects emit light. The wave theory predicted an absurd result at high frequencies, a prediction so wrong it was called the ultraviolet catastrophe. Planck found that if energy could only be exchanged in discrete chunks proportional to frequency, the measurements matched exactly. He introduced the proportionality constant that now bears his name: Planck's constant. But Planck regarded his idea as a mathematical trick. It was Einstein who recognized that the quanta were real. Together, their contributions form the cornerstone of quantum mechanics.

The Double-Slit Experiment and Wave-Particle Duality

We return now to the double-slit experiment in its modern, single-photon form, because it is the clearest possible demonstration of the central mystery of quantum mechanics. You already know what Young's original experiment showed: light passing through two slits produces an interference pattern, the signature of waves. But here is the part that makes physicists pause even after decades of working with it.

Suppose you reduce the light source to such low intensity that only one photon is in the apparatus at a time. Modern technology actually allows this. You fire individual photons one at a time, with a gap between them long enough that the previous photon has already been detected before the next one is released. Each photon travels alone, hits the detector screen, and leaves a single dot. The first dot appears at a random location. So does the second. And the third. After a hundred dots, you see nothing but scattered points.

But then you keep going. After a thousand dots, a faint structure begins to emerge. After ten thousand, the structure is unmistakable: you have an interference pattern, the same alternating bright and dark bands that Young saw with a continuous beam. Each photon landed at a random spot, but the ensemble of many photons organized itself into the interference pattern that only waves produce. No photon interacted with any other photon, because they traveled one at a time. Whatever produced the pattern, each photon was doing it alone, with itself.

Diagram 1.5 - The Double-Slit Experiment: One Photon at a Time

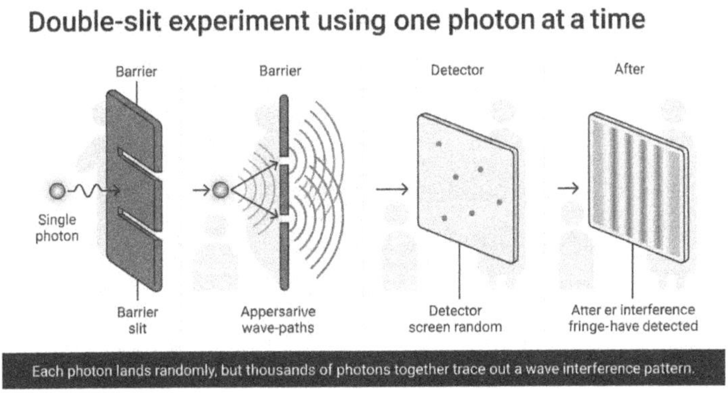

Double-slit experiment using one photon at a time

Each photon lands randomly, but thousands of photons together trace out a wave interference pattern.

The natural response is to reach for the most comfortable explanation: maybe the photon really does split in two, passes through both slits, and recombines at the detector. But experiment rules this out. If you place a detector at one of the slits to find out which slit the photon passed through, the interference pattern disappears. The moment you gain knowledge about the photon's path, the wave-like behavior vanishes. Turn off the detector, and the interference pattern comes back. The photon seems to 'know' whether it is being watched.

Quantum mechanics provides a resolution that is conceptually demanding but internally consistent. Before it is detected, a photon does not have a definite location or a definite path. It exists in a quantum state that includes all the possible paths it could take, each weighted by a probability. The wave you observe is not a physical wave made of matter: it is a probability wave, a mathematical description of where the photon is likely to be found when you look. When two probability waves overlap, they can interfere constructively or destructively. That is why some locations on the screen are far more likely to receive a photon than others. The moment you detect the photon, the probability wave collapses to a single definite location.

Diagram 1.6 - Probability Waves and Detection: The Collapse of Superposition

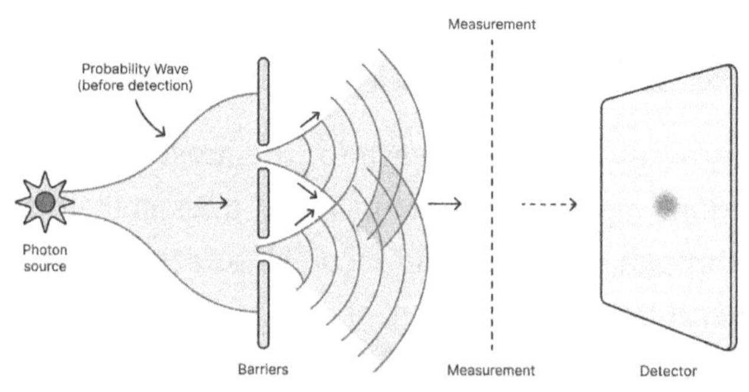

Before detection, the photon exists as a spread-out probability wave; the moment it is detected, it snaps to one location.

What a Photon Actually Is

At this point, it is worth pausing to build a clear and accurate picture of the photon, as the term is used loosely and confusingly. A photon is the fundamental quantum unit of the electromagnetic field. It is the smallest possible package of electromagnetic radiation, indivisible in the sense that it cannot be split into smaller packets of the same kind. When an atom releases energy by dropping from a higher energy level to a lower one, that energy is not released gradually; it is emitted as a single photon, all at once.

The photon is massless. It has no rest mass, which means it cannot be brought to rest in the ordinary sense. The moment it stops traveling, it ceases to exist as a photon: it has been absorbed. Because it is massless, it always travels at the speed of light in a vacuum, approximately 299,792 kilometers per second. This speed is not something photons achieve by being pushed hard enough. It is the only speed at which a massless particle can exist. Slow it down, and it disappears.

The energy a photon carries is determined entirely by its frequency, which is the rate at which its associated electromagnetic field oscillates. Higher frequency means higher energy. Visible light spans a range of frequencies, with violet at the high-frequency end and red at the low-frequency end. A single violet photon carries roughly twice the energy of a single red photon. This is why ultraviolet light can damage skin and eyes. In contrast, visible light does not: each UV photon carries enough energy to break chemical bonds in biological molecules, while a red-light photon does not, regardless of how many red photons are present.

Diagram 1.7 - A Photon's Properties: Frequency, Energy, and Polarization

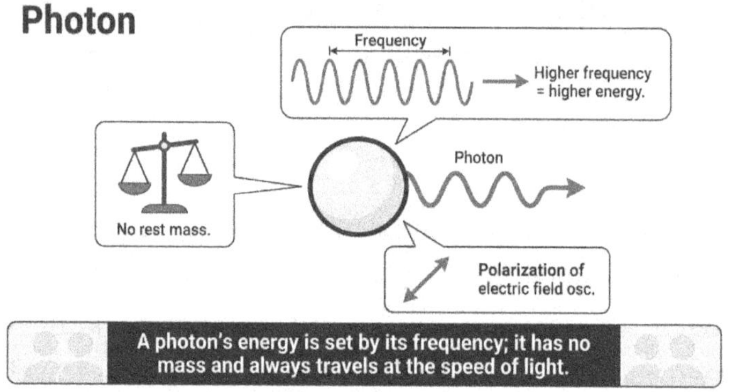

Photons also carry a property called polarization, which describes the orientation in which the electric field component oscillates as the photon travels. Think of it as the direction in which the wave wiggles as it moves forward. For a photon traveling toward you, the electric field might oscillate up and down (vertical polarization), left and right (horizontal polarization), or in a rotating corkscrew pattern (circular polarization). Polarization is not a minor technicality. It is one of the key properties exploited in quantum communication systems, where different polarization states encode the quantum information carried by each photon.

Photons also interact with matter by being absorbed and emitted whole, never in fractions. When an atom absorbs a photon, the atom gains the photon's entire packet of energy, and the photon ceases to exist. When the atom later releases that energy, it emits a new photon with the appropriate frequency. This all-or-nothing transaction gives rise to the sharp, precise spectral lines that astronomers use to identify the chemical composition of distant stars: each element absorbs and emits photons at very specific frequencies determined by its electron energy levels, producing a fingerprint in the light spectrum.

Wave-Particle Duality as a Feature, Not a Bug

It would be easy to feel that wave-particle duality is a problem with our understanding, a sign that physicists have not yet figured out what light really is. That feeling is backward. Wave-particle duality is not a confession of ignorance. It is a precise and well-tested description of how nature actually works at the quantum scale. The discomfort comes from trying to force quantum objects into the mold of classical objects, and classical objects are simply the wrong template.

Consider an analogy. Imagine you are trying to describe a living tree using only two categories: liquid or solid. The trunk is clearly solid, yet sap flows through it, and cells are mostly water. Is the tree a solid or a liquid? The dichotomy was never the right framework. The particle-versus-wave dichotomy is the same kind of wrong framework for describing a photon. A photon has particle-like properties in some experimental contexts and wave-like properties in others. The mathematical description that covers both is quantum mechanics.

Diagram 1.8 - Duality in Context: When the Wave Shows and When the Particle Shows

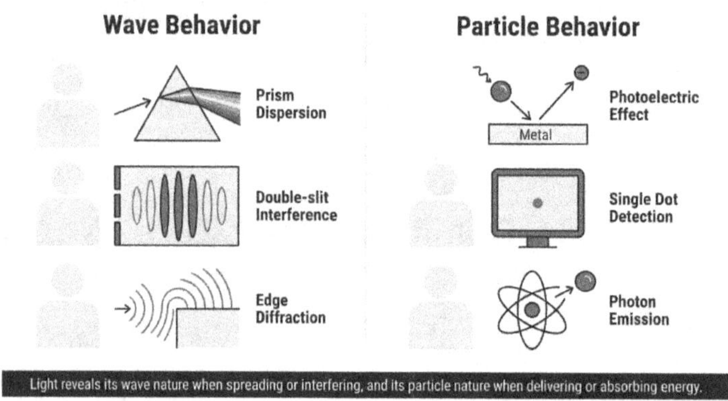

What makes duality not just acceptable but genuinely exciting is that each face of the dual nature is useful in technology. The wave nature of light makes interference-based sensing possible. Instruments called interferometers use the fact that photons travel as probability waves and can interfere with one another to measure distances with extraordinary precision. The gravitational-wave detectors at LIGO, which detected ripples in spacetime from colliding black holes, work on exactly this principle. They are interferometers of gigantic scale, and their sensitivity depends entirely on the wave nature of light.

The particle nature of light, on the other hand, is what makes quantum communication possible. The fact that a photon is indivisible and that measuring it inevitably disturbs it is the foundation of quantum key distribution. You cannot eavesdrop on a stream of single photons without leaving a detectable trace, because intercepting a photon means absorbing it, and absorbing it means changing what arrives at the intended recipient. The photon's particle nature is not an inconvenient limit: it is the security guarantee.

And then there is the superposition principle, perhaps the deepest consequence of wave-particle duality for quantum computing. Because a photon travels as a probability wave, it can be in a superposition of multiple states simultaneously until it is measured. A photon can be in a superposition of vertical and horizontal polarizations at the same time, meaning it is not definitely one or the other but rather some combination of both. When you use polarization states to encode quantum bits, called qubits, this superposition means a single photon can represent 0 and 1 simultaneously. That is the foundation of quantum parallel processing.

Diagram 1.9 - Superposition: A Photon Polarization Qubit

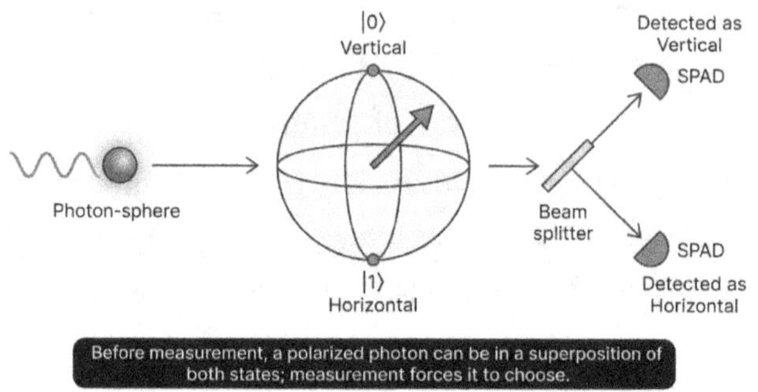

The Photon as the Foundation of All Photonics

Every technology in the chapters ahead is, at its core, a technology for working with photons: generating them, shaping them, guiding them, measuring them, and, in some cases, entangling them. Understanding what a photon is does not just help you appreciate those technologies abstractly: it helps you understand why they work the way they do, why they have the limits they have, and what it would take to push those limits further.

Consider the fiber-optic cable. At first glance, a fiber is just a very thin glass wire that carries light instead of electrical current. But the reason it can carry terabits of data per second across continents is rooted in the properties of photons. Photons do not lose energy to heat as they travel, unlike electrons in a conductor. They do not interact with each other, even when millions share the same fiber, which means you can send multiple colors of light through the same fiber simultaneously without interference between channels. And their speed is the speed of light, the universal maximum.

Lasers are another example. A laser produces photons in an extraordinarily organized way: all the same frequency, all traveling in the same direction, all with their fields oscillating in lockstep. This coherent behavior is possible because photons are bosons, a category of particles that can all occupy the same quantum state simultaneously. Electrons cannot do this; they crowd each other out by the Pauli exclusion principle. The surgical laser, the Blu-ray player, and the barcode scanner all exploit this bosonic nature of the photon.

Diagram 1.10 - Photons at Work: Five Technologies, One Particle

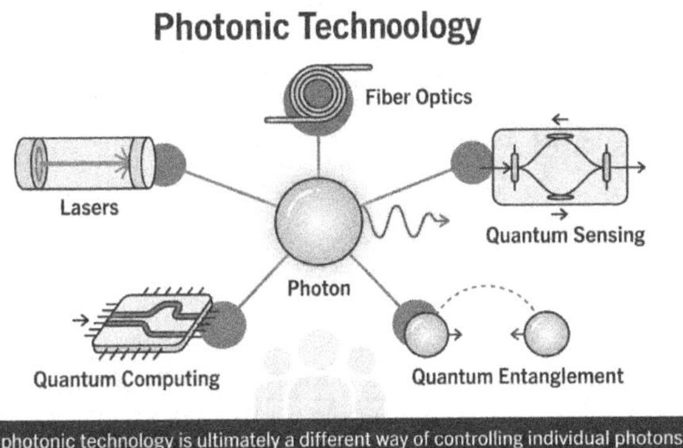

Quantum communication systems leverage the indivisibility of photons to provide a security guarantee. Because a photon cannot be split in two, an eavesdropper cannot tap a quantum channel the way you can tap an

electrical wire by diverting a fraction of the current. Intercepting a quantum signal requires catching individual photons, which inevitably disturbs the signal and alerts the intended parties to the intrusion. This guarantee is not a feature added by clever engineering: it is baked into the photon's fundamental nature.

Photonic quantum computing pushes even further. In a photonic quantum computer, qubits are encoded in properties of individual photons: polarization, path, or arrival time. Operations are performed by routing photons through beam splitters and phase shifters on tiny silicon chips. The superposition and entanglement properties of photons allow these chips to perform certain computations exponentially faster than any classical machine. The photon turns out to be an almost ideal carrier of quantum information.

Diagram 1.11 - From Atom to Channel: How a Photon Is Born, Travels, and Arrives

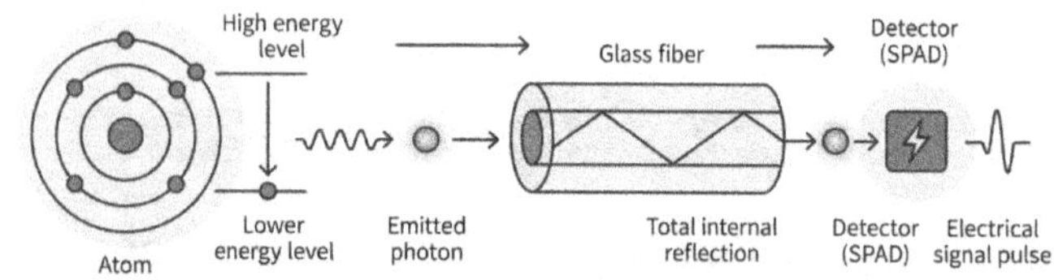

Every optical communication link begins with a photon born in an atom and ends with a photon to absorbed by a detector.

What This Means For You

If you are an engineer, a technologist, or a practitioner in any field that touches photonics, the concepts in this chapter are not historical curiosities. They are the active foundation of the systems you work with or will work with soon. The wave-particle duality of the photon determines how optical systems can be designed and what limits they cannot exceed. The quantized nature of photon energy determines the noise floors of optical detectors and the minimum signal levels that quantum communication systems must handle. The polarization properties of photons are the variables that quantum information systems encode and measure.

For those tracking the quantum technology landscape, whether assessing investment opportunities or briefing leadership, this chapter gives you the vocabulary to ask the right questions. When a company claims its quantum communication system is physically unbreakable, you now understand the mechanism: a single photon cannot be cloned or intercepted without detection. When a photonic computing startup mentions 'boson sampling,' you understand that bosons can occupy the same state simultaneously, creating a computationally hard problem that a photonic device handles naturally.

For the curious reader who came here to understand one of the most fascinating subjects in modern science, you now know that light is not a simple thing. It is not just a wave, and it is not just a particle. It is something far richer, a quantum entity that defies classical categories and obeys rules that are strange, precise, and deeply beautiful. The phone flashlight that lit up your room last night was emitting trillions of photons per second, each a quantum of the electromagnetic field, each traveling at the universal speed limit, each interacting with atoms in your room as a precise packet of energy. Ordinary life is built on quantum mechanics, even when nobody notices.

Takeaway

Light has a history of surprising the people who studied it most carefully. Newton's particles explained reflection but not interference. Huygens's waves explained interference but not the photoelectric effect. Maxwell's electromagnetic synthesis unified two great forces of nature and predicted the speed of light from pure mathematics. Einstein's photon resolved the ultraviolet catastrophe and the photoelectric anomaly by showing that light comes in indivisible packets. And the double-slit experiment carried out with one photon at a time revealed that each of those packets travels through space as a probability wave that can interfere with itself. Every upgrade came from taking the experimental evidence seriously, even when it was uncomfortable.

The photon that emerged from this long debate is a genuinely extraordinary entity. It is massless, so it travels at the speed of light by necessity. It carries a precisely quantized packet of energy determined by its frequency. It has polarization, which can be used to encode information. It is a boson, so many photons can occupy the same state, making lasers possible. It cannot be divided or copied without detection, making quantum cryptography possible. And it travels as a probability wave that can be put into superposition and made to interfere, enabling quantum sensing and quantum computing.

Carry this picture forward. Each chapter in this book will introduce new phenomena, technologies, and experimental results. Still, they will all be stories about what happens when you generate photons precisely, guide them carefully, and sometimes entangle them with each other. The more vividly you can picture a photon, that massless, wave-like, particle-like, polarized, indivisible quantum of light, the more deeply you will understand what comes next. You are not just learning about light. You are building the conceptual vocabulary for one of the most consequential technological fields of the twenty-first century.

Diagram 1.12 - The Photon's Conceptual Map: Building Blocks of the Book Ahead

Infographic Roadmap

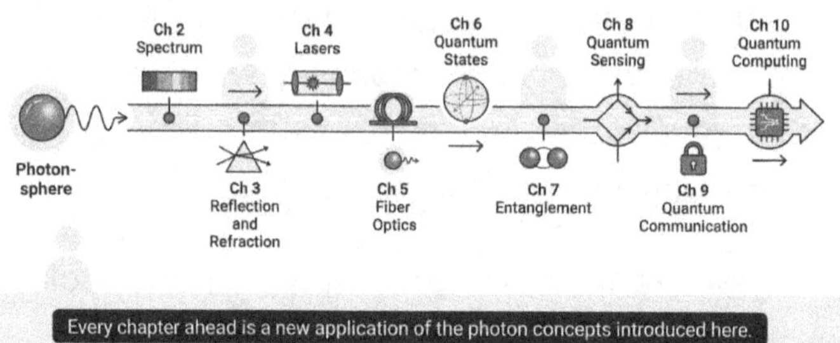

The sunbeam through your window and the flashlight on your phone are made of the same particles, governed by the same rules, as the photons in a quantum key distribution system or a photonic quantum computer. The difference is only in how precisely those particles have been prepared, controlled, and measured. That precision is what photonics is about, and it begins with understanding what a photon is.

2 The Electromagnetic Spectrum: Light's Full Range

Opening Scenario

The morning starts with a routine hospital check-in. A staff member points a small handheld device at your forehead, and in two seconds, a number appears: your body temperature, measured without a single point of contact. That gadget emits invisible infrared radiation onto your skin and reads the heat your body already radiates. You do not feel a thing. But a specific slice of the electromagnetic spectrum just traveled from your forehead to a detector the size of a thumb drive and turned body heat into a precise number.

Down the corridor, a patient slides into the wide white tunnel of an MRI scanner. The machine uses radio waves, the longest and lowest-energy members of the electromagnetic spectrum, to interact with hydrogen atoms packed into human tissue, building a three-dimensional map of soft structures that would have seemed miraculous a century ago. In the next room, an X-ray technician drapes a lead apron over a patient's torso. A burst of high-energy photons punches through skin and muscle, casting shadows of bone onto a detector. Across the hall, a UV sterilization lamp floods an empty procedure room for forty-five seconds, destroying the genetic material of every bacterium and virus on every surface.

All of this unfolds inside a single building on an ordinary morning. Every technology involved uses a different band of the same extended family of radiation that includes the light your eyes can see. These are the quiet instruments of modern life, each tuned to a frequency range your eyes cannot reach. This chapter is your map of that family: what its members are, how they relate to one another, and why the physical rules governing each band determine what it can and cannot do.

Why It Matters

Photonics focuses on a relatively narrow portion of the electromagnetic spectrum: visible light, the near-infrared band used in fiber-optic cables, and the ultraviolet frequencies used in chip manufacturing. But you cannot appreciate why photonics concentrates there unless you understand the broader landscape. The spectrum is not a collection of unrelated phenomena. It is a single, continuous physical reality governed by identical principles from end to end. Radio waves and gamma rays obey the same fundamental physics; the differences between them are purely differences of frequency, wavelength, and photon energy. Those energy differences are what make some regions gentle and others lethal, some ideal for communication and others for imaging bone.

For engineers and practitioners, the spectrum provides critical context. The near-infrared frequencies favored by fiber optic systems were not arbitrary choices. They were selected because silica glass transmits those particular wavelengths with exceptionally low loss. By the time you finish this chapter, the design choices made by photonics engineers will look less like arbitrary decisions and more like logical consequences of physical reality.

One Family, Many Faces: The Spectrum as a Unified Whole

It is tempting to think of radio waves, X-rays, and visible light as fundamentally different things. Radio waves carry music to your car without harming you. X-rays require lead shielding. Visible light lets you read a book. Yet these are not three different physical phenomena. They are the same phenomenon at vastly different frequencies, in the same way that a whisper and a jet engine are both pressure waves in air, just at vastly different energy levels.

The unifying concept is electromagnetic radiation. Every form of it consists of two oscillating fields, an electric field and a magnetic field, vibrating at right angles to each other and to the direction of travel. These fields require no physical medium. They propagate through the complete vacuum of outer space, which is why sunlight can cross 93 million miles of emptiness and warm your face on a spring morning.

The physicist James Clerk Maxwell unified electricity, magnetism, and light in the 1860s. He showed that a changing electric field generates a magnetic field, and vice versa, creating a self-sustaining wave that travels at the speed of light. Maxwell realized he had discovered what light actually was. When your Wi-Fi router broadcasts a signal and when a nuclear reactor emits gamma rays, both processes obey the same equations Maxwell wrote more than 150 years ago. What separates these two is simply frequency: how many times per second the electromagnetic field completes a full oscillation cycle. AM radio oscillates about one million times per second; a visible green photon oscillates about 600 trillion times; a hard X-ray at frequencies in the quintillions. This enormous range, spanning more than twenty-five powers of ten, is the electromagnetic spectrum. It's named regions, radio through gamma ray, are human labels placed at convenient points along a continuous gradient, much as we label sections of a rainbow, even though no sharp line separates orange from yellow.

Diagram 2.1 - The Electromagnetic Spectrum: One Continuous Family

Full Electromagnetic Spectrum

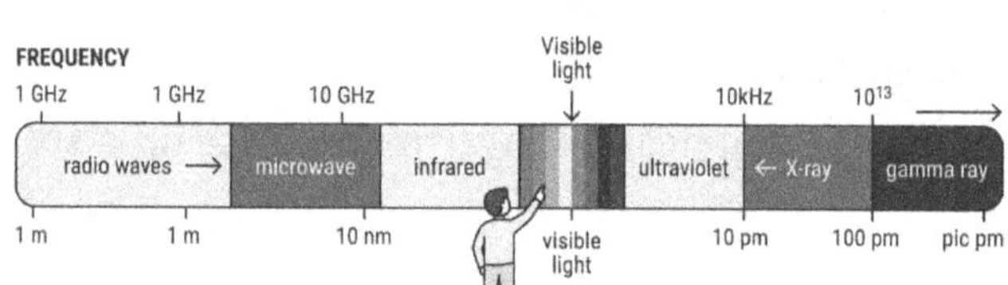

All electromagnetic radiation is the same phenomenon at different frequencies; only the scale changes.

Because all spectral regions share the same physics, advances in one area often illuminate possibilities in another. Electromagnetic radiation can be absorbed by matter, reflected off a surface, transmitted through a material, or scattered by small structures. These four interaction types apply across the entire spectrum; only the specific rules governing each depend on frequency and material properties.

Frequency, Wavelength, and Energy: The Three-Way Relationship

Three quantities characterize any electromagnetic wave: its frequency, its wavelength, and the energy carried by each of its photons. These three are not independent. They are locked together in a relationship that, once you understand it conceptually, makes the entire spectrum far easier to navigate.

Frequency is the rate of oscillation: how many complete wave cycles pass a fixed point every second, measured in hertz. Visible light lies in the hundreds-of-terahertz range, meaning hundreds of trillions of cycles per second. AM radio broadcasts sit in the megahertz range, millions of cycles per second. Wavelength is the physical size of one complete cycle: the distance from one wave crest to the next. Because all electromagnetic waves travel at the same speed in a vacuum, frequency and wavelength are inversely related. High-frequency waves have shorter physical cycles; low-frequency waves have longer ones. Radio waves can have wavelengths ranging from meters to kilometers. Gamma rays have wavelengths smaller than an atomic nucleus. Visible light sits between roughly 380 and 700 nanometers, about 100 times thinner than a human hair.

Diagram 2.2 - Frequency, Wavelength, and Photon Energy: The Three-Way Link

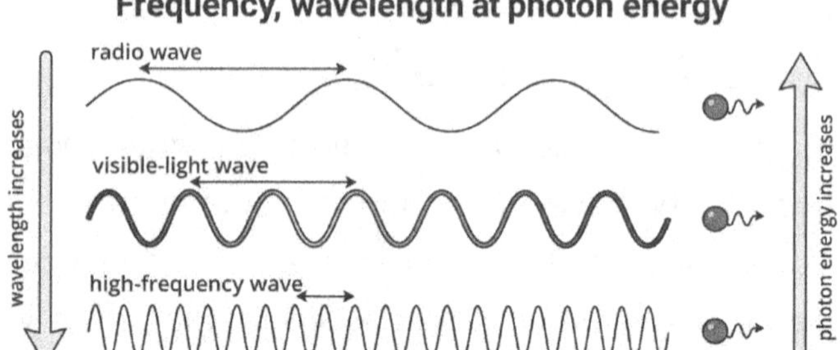

Higher frequency means shorter wavelength and more energy per photon.

The third quantity, photon energy, connects quantum physics to wave physics. Each photon carries energy directly proportional to its frequency: double the frequency, and you double the energy per photon. This proportionality is governed by Planck's constant, named for Max Planck, who first recognized that light comes in discrete energy chunks. The key implication is simple: higher-frequency photons are more energetic, and that extra energy determines how aggressively a photon interacts with matter.

This three-way relationship governs every electromagnetic application. Radio waves are safe to broadcast through living tissue because their photons carry too little energy to break chemical bonds. Gamma rays require lead shielding because their photons ionize atoms and disrupt DNA. UV causes sunburn while visible light from the same sun does not, because UV photons carry just enough extra energy to damage skin cell molecules while visible-light photons fall slightly short of that threshold. Every safety protocol and material choice in electromagnetic technology flows from this three-way relationship.

Why Human Eyes See Exactly What They See

Of all available frequencies, the human eye responds only to wavelengths between roughly 380 and 700 nanometers, a window we call visible light. This spans less than one octave of the full spectral range. Why this particular window? The answer is evolutionary, and it reveals something beautiful about the relationship between life and physics.

The Sun's surface temperature causes it to radiate most intensely at wavelengths clustered around the middle of the visible range, peaking near yellow-green. Earth's atmosphere is transparent in this visible window but strongly absorbs at many infrared and ultraviolet wavelengths. So the visible band is also the frequency range that most abundantly reaches Earth's surface. Life evolved under solar illumination in exactly this photon bath. Any organism that could detect these photons gained a decisive advantage for navigating, hunting, avoiding predators, and finding mates. Four billion years of natural selection built a biological photon detector optimized for solar output filtered through our atmosphere.

Diagram 2.3 - The Sun's Output and the Human Visual Window

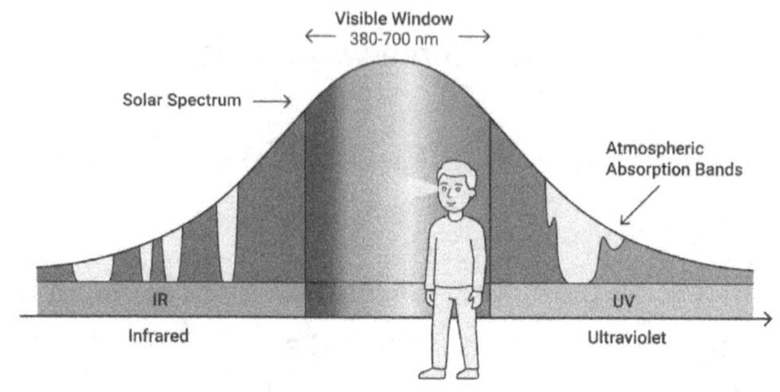

Human eyes evolved to detect the exact wavelengths the Sun produces most abundantly at Earth's surface.

The three types of color-sensitive cells in your retina, called cone cells, are tuned to long (reds), medium (greens), and short (blues) wavelengths. Your brain combines these signals to construct the full-color experience of a flower garden or a traffic light. Color is the brain's interpretation of the frequency distribution of photons at the retina. Some animals evolved visual systems tuned elsewhere: bees see ultraviolet patterns in flower petals invisible to us, while pit vipers detect infrared radiation from prey's body heat. The visible spectrum is not a universal constant but a window evolution opened for one species under one set of lighting conditions. The rest of the spectrum was always present; humans needed instruments to access it.

This evolutionary story connects directly to photonics engineering. Human eyes are not well-matched to the near-infrared wavelengths used in fiber-optic communications. A fiber carrying a terabit of data per second is invisible to your eye: those photons do not activate cone or rod cells. The detectors used in fiber optic receivers are semiconductor devices specifically engineered to respond to near-infrared photons. Every photonic instrument is essentially a new sensory organ, tuned not to the solar frequencies that shaped the retina but to whatever spectral band a technology requires.

Infrared: The Workhorse of Photonics and Thermal Imaging

Infrared radiation sits just beyond the red end of the visible spectrum, at wavelengths longer than about 700 nanometers. The word infrared literally means 'below red,' referring to its position below visible red light on the frequency scale. It spans a broad range divided into three sub-regions: near-infrared, mid-infrared, and far-infrared (also called thermal infrared). These sub-regions differ enough to support distinct technologies.

Near-infrared, from roughly 700 to 2500 nanometers, is the backbone of modern fiber optic communication. Silica glass transmits near-infrared light with exceptionally low loss: a signal can travel many kilometers through high-quality fiber before it fades significantly. At visible wavelengths, the same glass absorbs more light, dramatically shortening the signal's range. Two near-infrared windows in particular, around 1310 nanometers and 1550 nanometers, correspond to valleys in the absorption spectrum of silica glass where loss is minimized. These wavelengths were not arbitrary choices; they were selected because the physics of glass demanded them. Chapter 5 will explore fiber optic systems in depth.

For now, the key point is that the interaction between near-infrared light and glass enables long-distance optical communication.

Diagram 2.4 - Near-Infrared and the Fiber Optic Sweet Spot

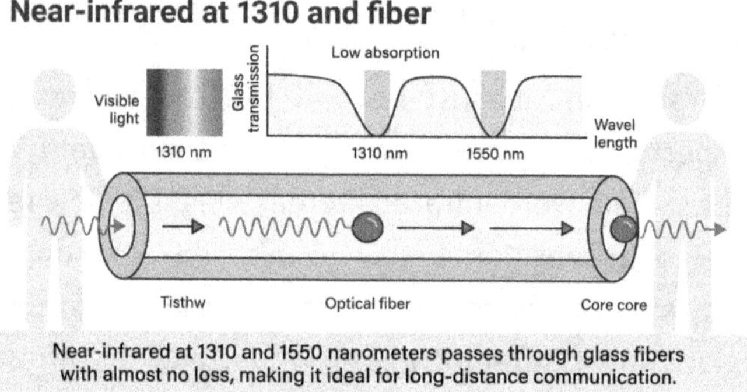

Near-infrared at 1310 and 1550 nanometers passes through glass fibers with almost no loss, making it ideal for long-distance communication.

Near-infrared also powers everyday sensing and remote control. The television remote in most homes emits pulses of near-infrared light invisible to the eye but detectable by a smartphone camera. Security systems, medical diagnostics, and facial recognition cameras all exploit structured near-infrared patterns. This invisible light does real work constantly, in every room you occupy.

Thermal infrared, at wavelengths roughly between 8 and 14 micrometers, is emitted by objects at everyday temperatures, including the human body. Every object warmer than absolute zero emits electromagnetic radiation, and for objects near room or body temperature, this emission falls squarely in the thermal infrared range.

The handheld thermometer at the hospital entrance was detecting precisely this: the infrared radiation naturally emitted by warm skin. No separate infrared source is needed. The detector measures the intensity and wavelength distribution of radiation arriving from your forehead and converts that into a temperature reading.

Diagram 2.5 - Thermal Infrared Imaging: Reading Heat as Light

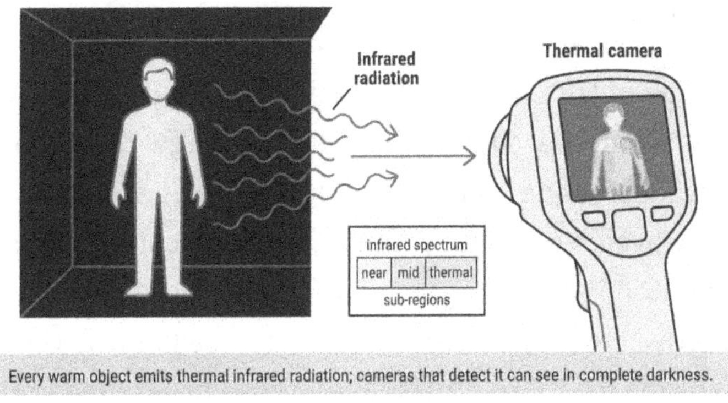

Every warm object emits thermal infrared radiation; cameras that detect it can see in complete darkness.

Thermal infrared cameras require no external light source because every object in their field of view emits infrared radiation proportional to its temperature. Building inspectors use thermograms to find heat loss through poorly insulated walls. Electrical engineers use them to detect overheating components before failure occurs. Search-and-rescue teams locate people in darkness or smoke. In every case, the camera is not projecting light; it is reading the invisible radiation that the scene is already emitting.

Ultraviolet: Sterilization, Lithography, and the Chip Revolution

Ultraviolet light lies just above visible violet, with wavelengths ranging from roughly 10 to 400 nanometers, subdivided into near-UV (UVA), mid-UV (UVB), far-UV (UVC), and deep- and extreme-UV bands used in chip manufacturing. What unifies all of ultraviolet is that its photons carry enough energy to interact with chemical bonds, enabling them to drive reactions that visible-light photons cannot.

The energy of UV photons is both a hazard and a tool. UVB photons from the sun carry enough energy to damage DNA strands inside skin cells. DNA absorbs UV light at specific wavelengths, and when a DNA molecule absorbs a UVB photon, it can cause the two strands to fuse incorrectly, creating a structural error. Cells have repair mechanisms for these errors, but intense or prolonged UV exposure creates errors faster than repairs can keep up, eventually leading to the mutations associated with skin cancer. Sunscreen works by absorbing UV photons before they reach skin cells and converting their energy into harmless heat.

That same DNA-damaging property becomes a powerful tool in UVC sterilization. The DNA and RNA of bacteria and viruses efficiently absorb UVC light at wavelengths around 254 nanometers. When a microorganism's genetic material absorbs a sufficient dose of UVC photons, it accumulates enough structural damage that the organism can no longer replicate. UV sterilization systems now disinfect hospital rooms, water treatment facilities, air-handling systems, and food-processing equipment without the need for chemical disinfectants.

Diagram 2.6 - UV Sterilization: Photons as Disinfectants

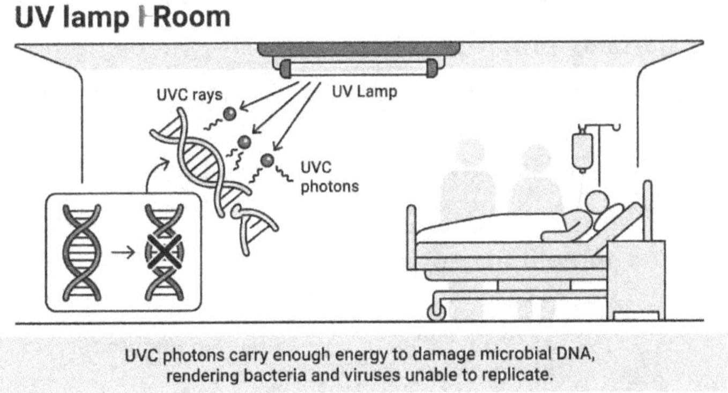

UVC photons carry enough energy to damage microbial DNA, rendering bacteria and viruses unable to replicate.

The most economically consequential application of ultraviolet light is photolithography, the process by which patterns are printed onto silicon wafers to create microprocessors and memory chips. The core principle is simple: the minimum feature size you can print with light is limited by the wavelength of that light. Longer wavelengths produce blurrier patterns with larger minimum features. Shorter wavelengths produce sharper patterns with smaller minimum features.

Early integrated circuits were made using visible light, with feature sizes measured in micrometers. As chip makers pushed toward smaller transistors, they shifted to near-UV, then deep-UV, and finally extreme-UV wavelengths.

Today's leading-edge chips, with transistors a few nanometers in size, are manufactured using extreme-UV lithography at wavelengths of just 13.5 nanometers. The machines producing this extreme-UV light are among the most complex industrial instruments ever built, each costing several hundred million dollars. The entire modern semiconductor industry exists because engineers learned how to harness progressively shorter UV wavelengths to draw progressively finer patterns on silicon.

Diagram 2.7 - UV Lithography: Printing Circuits with Light

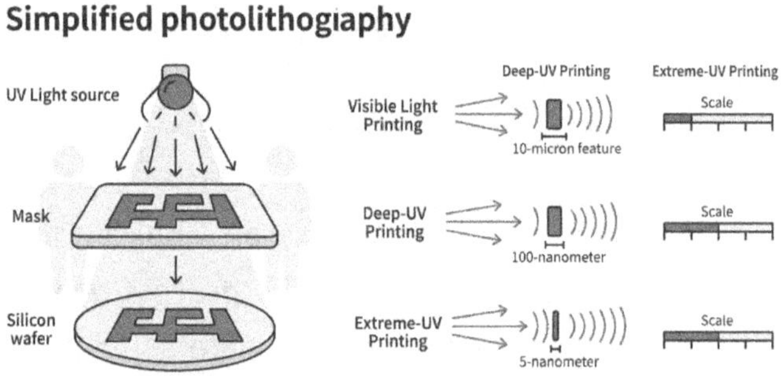

Shorter ultraviolet wavelengths allow smaller circuit features: the entire history of Moores Law is written in shrinking wavelengths.

Terahertz: Bridging the Gap Between Electronics and Photonics

Between the microwave and infrared regions lies a frequency band once called the terahertz gap, not because the band was unimportant, but because it was genuinely difficult to generate and detect. The terahertz band spans roughly 100 gigahertz to 10 terahertz. Traditional electronic oscillators used in radar and wireless systems struggle to reach these frequencies. Traditional photonic sources, such as lasers, typically produce much higher frequencies. For most of the twentieth century, terahertz radiation occupied an awkward no-man's-land between the two great domains of electromagnetic technology.

Advances in ultrafast laser technology and semiconductor physics over the past three decades have closed this gap. Researchers discovered that when a femtosecond laser pulse (lasting a few millionths of a billionth of a second) strikes certain semiconductor materials, it generates a burst of terahertz radiation. Other approaches use specially designed semiconductor devices that oscillate at terahertz frequencies, or exploit nonlinear optical effects in crystals to convert near-infrared laser light into terahertz radiation. These methods remain more expensive than generating microwaves or visible light, but they have made terahertz technology practical for important applications.

Diagram 2.8 - The Terahertz Gap: Between Electronics and Photonics

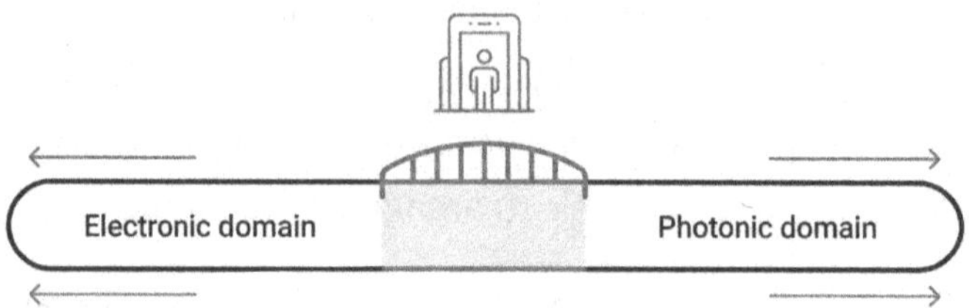

The terahertz band was long a technological gap; advances in ultrafast lasers have made it accessible for imaging and sensing.

The opening scenario of this chapter illustrated the most visible application of terahertz technology. Common non-conducting materials, including fabric, paper, cardboard, plastic, leather, and dry wood, are largely transparent to terahertz radiation. Metals, ceramics, certain explosives, and biological tissue reflect or absorb it in distinct ways. This contrast allows a terahertz scanner to see through clothing and map objects beneath it, producing detailed images of hidden objects that security officers examine at airport checkpoints. The scanner does not pass radiation through the entire body the way X-rays do, which limits its depth reach but eliminates exposure to ionizing radiation.

Beyond security, terahertz spectroscopy exploits the fact that many molecules absorb terahertz radiation at specific frequencies corresponding to their rotation and collective vibration modes. These spectral fingerprints identify molecular species with great specificity. Pharmaceutical manufacturers use terahertz spectroscopy to verify the chemical identity and structural integrity of tablets through their packaging, speeding quality control without destroying samples. Customs agencies use it to screen sealed packages. Art conservators use it to examine paint and varnish layers without removing material. The same physics underlies all of these applications: the interaction of terahertz photons with molecular structure at the appropriate frequency.

There is also growing interest in terahertz wireless communication. As demand for wireless data capacity grows, engineers push into progressively higher frequency bands because higher frequencies can carry more data per second. The millimeter-wave bands used in 5G networks sit just below the terahertz boundary. True terahertz wireless links, with data rates measured in terabits per second, are an active area of research. The challenges are real: terahertz signals are absorbed by atmospheric water vapor and do not travel well through walls. But for short-range, high-capacity links inside data centers or between adjacent devices, terahertz communication is a credible frontier.

X-rays and Gamma Rays: Probing Matter's Deep Structure

At the highest-energy end of the spectrum sit X-rays and gamma rays, photons energetic enough to ionize atoms, penetrate dense matter, and interact with atomic nuclei. The distinction between X-rays and gamma rays is partly historical and partly based on origin: X-rays are typically produced by interactions involving electrons in atoms or the deceleration of high-speed electrons, while nuclear processes produce gamma rays. Physically, a high-energy X-ray photon and a low-energy gamma-ray photon are indistinguishable. The label depends on how the photon was produced, not what it is.

Medical X-ray imaging exploits straightforward physics. An X-ray machine accelerates electrons to high speed and fires them at a metal target. When these fast electrons are abruptly decelerated, they release kinetic energy as X-ray photons. Some of these photons pass through the patient's body with minimal interaction; others are absorbed by dense tissues, such as bone, which have more atoms per unit volume. A detector on the far side captures the transmitted photons, producing a shadow map in which dense tissue appears lighter (more photons blocked) and soft tissue appears darker (more photons transmitted). A chest X-ray is literally a density map of your internal structures, cast by photons.

Diagram 2.9 - Medical X-ray Imaging: Shadow Maps of Internal Density

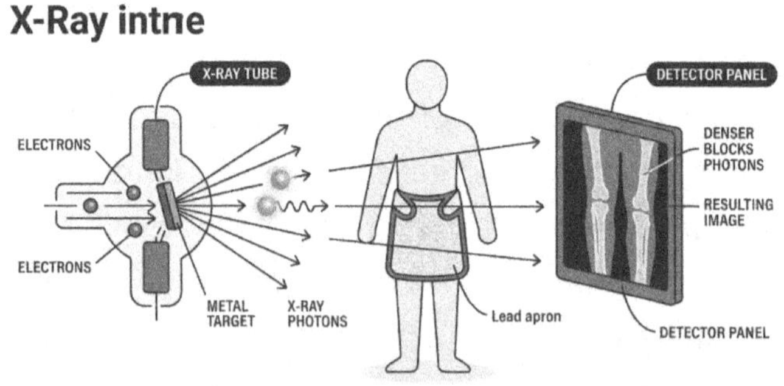

An X-ray image is a shadow map: dense bone blocks more photons, appearing lighter; soft tissue transmits more, appearing darker.

CT scanning extends X-ray imaging into three dimensions by acquiring hundreds of images from different angles and reconstructing a navigable internal model without a single incision. PET scanning goes further: a mildly radioactive tracer injected into the patient emits gamma rays as it decays, and the scanner maps their origins to reveal where the tracer accumulates in metabolically active tissue, thereby locating tumors or mapping brain activity. Synchrotron light sources, large ring-shaped particle accelerators, deliver X-ray brightness millions to billions of times higher than that of medical tubes, enabling X-ray crystallography that reveals the precise three-dimensional atomic arrangement of protein molecules. This capability has driven drug discovery for HIV, cancer, and influenza.

Diagram 2.10 - Synchrotron X-ray Source: Illuminating Molecules at the Atomic Scale

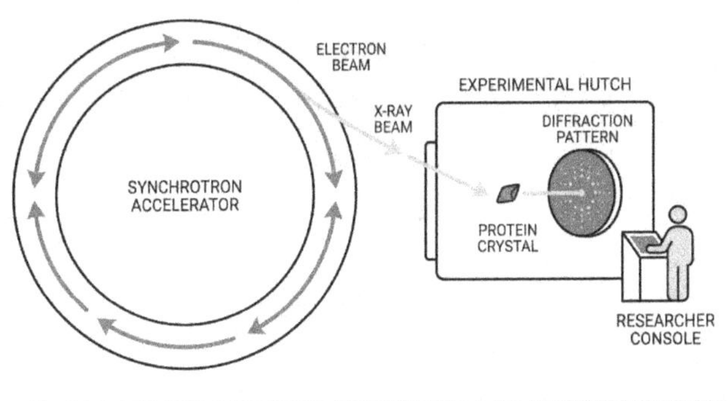

Synchrotron X-ray beams are billions of times brighter than medical tubes and can reveal the atomic structure of protein molecules.

Gamma rays present both extraordinary challenges and opportunities. Industrial scanning uses sealed radioactive sources to detect hidden flaws in welds and pipelines without disassembly. Food irradiation kills bacteria and inhibits sprouting to extend shelf life. In cancer treatment, stereotactic radiosurgery focuses gamma ray beams from multiple angles onto tumor tissue, delivering a high dose precisely to the target while sparing surrounding healthy tissue.

What This Means For You

If you work in engineering, technology, medicine, or any field involving sensing, communication, or materials analysis, this spectral map is a practical tool, not just abstract knowledge. Engineers designing wireless systems need to understand how frequency choice determines penetration through walls, atmospheric absorption, and achievable data rate. Biomedical device professionals need to understand why a thermal imaging camera and an X-ray machine both reveal internal structure but in fundamentally different ways. Software engineers building applications for depth cameras or AR headsets need to understand why those devices use near-infrared structured light rather than visible light. In every case, the physical logic of the spectrum underlies the design choice.

Diagram 2.11 - Spectral Bands and Their Primary Engineering Applications

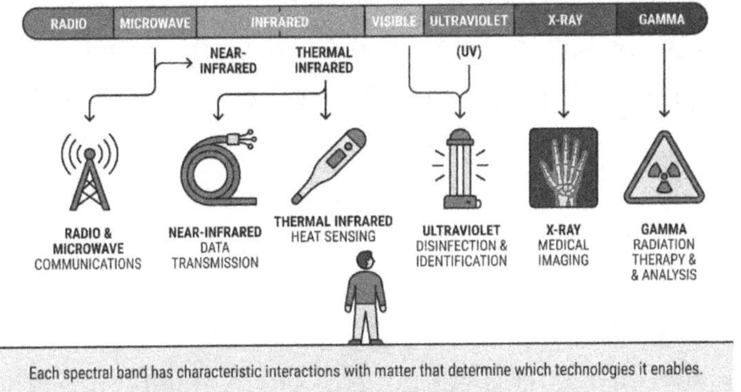

Each spectral band has characteristic interactions with matter that determine which technologies it enables.

For practitioners who use photonic instruments without designing them, spectral literacy sharpens both the interpretation of results and the critical evaluation of technology claims. When a manufacturer advertises a terahertz scanner as a replacement for X-ray inspection, you can ask the right questions: what materials need to be penetrated, what is the required depth, and does terahertz radiation's limited penetration of dense tissue matter for this application? When a colleague proposes visible-light cameras for a quality-control application to detect heat-related defects, you can recognize that a thermal infrared camera might be the more direct and informative tool.

The quantum dimension adds further relevance for those in emerging technology fields. Quantum communication uses single photons at near-infrared wavelengths in optical fibers. Photonic quantum computing uses photons as qubits, the fundamental unit of quantum information, at visible or near-infrared wavelengths. Quantum sensing achieves measurement sensitivity beyond classical limits by exploiting engineered quantum states of light at specific wavelengths. In every case, the choice of wavelength reflects the practical constraints of available materials, detectors, and light sources. The spectral map you have built in this chapter will be an indispensable context for every quantum technology discussed later.

Takeaway

The electromagnetic spectrum is a single, unified physical phenomenon spanning more than twenty-five orders of magnitude in frequency. All electromagnetic radiation, from radio waves that wrap around a building to gamma rays that penetrate steel, consists of oscillating electric and magnetic fields traveling at the speed of light. The named regions, from radio to gamma rays, are not separate phenomena but different octaves of the same instrument, differing only in frequency, wavelength, and the energy carried by each photon.

The three-way relationship among frequency, wavelength, and photon energy is the master key. Higher frequency means shorter wavelength and more energetic photons. More energetic photons interact more aggressively with matter. That is why gamma rays require lead shielding, and radio waves pass through walls harmlessly. Human eyes evolved to detect the visible band because it corresponds to the peak solar output at Earth's surface. Every other part of the spectrum was always present; we needed instruments to access it.

Photonics focuses on visible light, near-infrared, and ultraviolet wavelengths. Near-infrared is the workhorse of fiber optic communication because silica glass transmits it with minimal loss. Ultraviolet drives semiconductor manufacturing because shorter wavelengths allow finer printed features. Terahertz bridges electronics and photonics for security scanning and chemical identification. X-rays and gamma rays probe matter at atomic and nuclear scales, powering medicine, structural biology, and industrial inspection.

Diagram 2.12 - The Spectrum's Full Map: A Reader's Reference

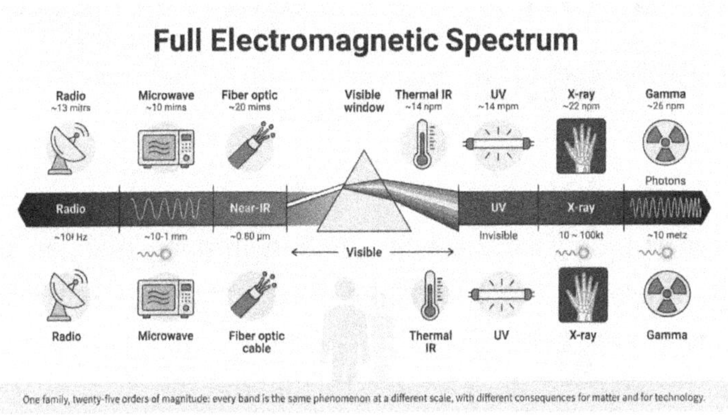

One family, twenty-five orders of magnitude: every band is the same phenomenon at a different scale, with different consequences for matter and for technology.

The hospital visit at the start of this chapter was not a collection of unrelated technologies housed in the same building. It was a demonstration that modern medicine, like modern engineering and modern communication, has learned to exploit the entire electromagnetic family, not just the narrow sliver our eyes can see. The infrared thermometer, the MRI scanner's radio waves, the X-ray machine, and the UV sterilization lamp are all instruments tuned to different bands of the same spectrum, operating by the same physics at different scales. Understanding their unity is the conceptual foundation for everything that follows in this book. Every laser, every fiber, every quantum communication protocol, and every photonic computing device is built on a carefully chosen location within this vast electromagnetic landscape. You now have the map.

3 How Light Behaves: Reflection, Refraction, and Diffraction

Opening Scenario

You are standing at the edge of a swimming pool on a bright afternoon. Look straight down, and you see the bottom clearly. Look toward the shallow end from across the pool, and the floor seems to tilt, the depth appears shallower than it is, and a swimmer's legs bend sharply at the waterline as though they belong to two different bodies. The feet seem closer and shifted sideways. You know the swimmer's legs are perfectly straight. But the path light takes from those submerged legs to your eyes has been bent at the water's surface, and the visual system, which assumes light always travels in straight lines, reconstructs an image that puts the legs where they are not. This is why a spear fisherman must aim below where a fish appears to be: the fish's image is displaced from the fish itself.

Now walk to the parking lot and look at the puddle near the entrance. Its surface is reflecting the sky with near-photographic fidelity: the blue overhead, a passing cloud, the silhouette of a nearby tree. Crouch closer and you see a thin film of oil from a leaking engine has spread across part of the puddle, turning the reflection into swirling bands of iridescent green, purple, and gold that shift as you tilt your head. The puddle is demonstrating two distinct behaviors simultaneously. The oil-free section bounces light back like a mirror, preserving angles and image detail. The oil-coated section splits incoming light into reflections from the film's top and bottom surfaces, and those two reflections interfere: some wavelengths cancel, others amplify, producing the color bands.

These two observations, the bent swimmer's legs and the oil-film rainbow, are direct demonstrations of refraction and interference, two of the three foundational behaviors of light that this chapter examines. Add reflection, visible on the undisturbed part of that same puddle, and you have the complete classical toolkit. Reflection, refraction, and diffraction are the operating principles behind every lens ever ground, every optical fiber ever laid, and every spectrometer that has probed the chemistry of a distant star. They are not textbook abstractions. They are the architecture of photonic engineering.

Why It Matters

The behaviors described in this chapter were first systematically characterized centuries ago, yet they remain the engineering foundation of photonic systems being designed and deployed right now. The same refraction that bends a swimmer's legs at the pool surface guides laser pulses through hair-thin glass fibers laid beneath the Atlantic Ocean, carrying hundreds of terabits of data per second between continents. The same interference that paints colors on an oil slick underlies the thin-film anti-reflection coatings on every camera lens, the wavelength-selective filters in optical telecommunications equipment, and the sensors that detect gravitational waves from merging black holes. The same diffraction that blurs an image when you stop a camera aperture too far down defines the fundamental resolution limit of every microscope, telescope, and semiconductor lithography machine on the planet.

For engineers and scientists working with light, these are not historical curiosities. They are the rules of the game. Every optical design decision, from selecting a glass type for a lens element to specifying a fiber splice angle to choosing a grating groove density for a spectrometer, is an application of the principles covered in this chapter. And for readers approaching quantum optics, where light's behavior at boundaries and apertures takes on stranger and more subtle character, a grounding in classical optical behavior is the map that makes quantum territory navigable. The quantum phenomena explored in later chapters will be far more surprising and far more meaningful when you understand what they depart from.

Reflection: Light Bouncing at Interfaces

When a beam of light traveling through air encounters a smooth surface, part of that light bounces back. This is a reflection. The defining rule is geometrically precise: the angle at which light arrives at a surface equals the angle at which it departs. Physicists express this by drawing an imaginary line called the normal, which runs perpendicular to the surface at the point of contact, and measuring angles relative to that line. The angle between the incoming ray and the normal is the angle of incidence. The angle between the outgoing reflected ray and the normal is the angle of reflection. These two angles are always equal. This relationship is the law of reflection and holds for every smooth reflective interface, from a bathroom mirror to the polished end face of a laser cavity.

What makes a surface smooth enough to produce this clean, angle-preserving reflection? The answer depends on the scale of surface roughness compared to the wavelength of the light. If surface irregularities are much smaller than the wavelength, the surface behaves optically smooth and produces specular reflection: the reflected rays preserve the spatial relationships of the incoming beam, producing a sharp image. If the surface roughness is comparable to or larger than the wavelength, the surface scatters light in all directions. This is diffuse reflection, which is why matte paper and rough concrete appear dull from every viewing angle rather than mirror-like from only one angle.

Diagram 3.1 - The Law of Reflection at a Mirror Surface

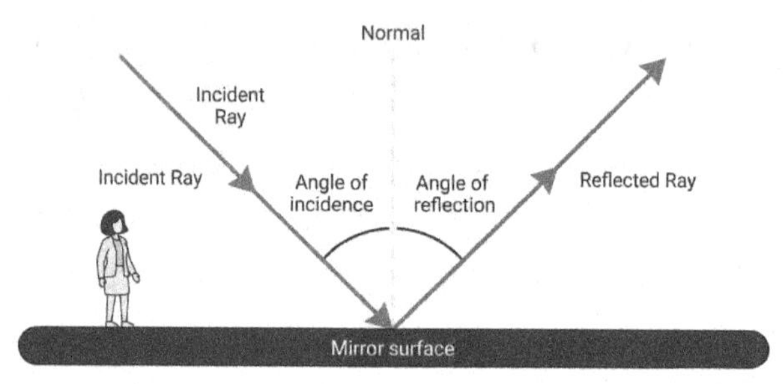

The angle of incidence always equals the angle of reflection, regardless of surface material.

Reflection does not require a metal mirror. Any interface between two materials of different refractive indices, a property explained in the next section, produces partial reflection. A glass window reflects roughly 4% of the light that hits it while transmitting the rest. Reflectivity rises sharply as the angle of incidence increases toward grazing angles, which is why a window becomes mirror-like when you look at it from the side rather than head-on. Optical engineers exploit this behavior deliberately. Anti-reflection coatings on camera lenses and eyeglasses are thin films designed so that the small reflections from the top and bottom surfaces of the film cancel each other through destructive interference, reducing reflectivity from 4% to a fraction of a percent. Highly reflective coatings stack multiple thin films so that reflections constructively reinforce, pushing reflectivity above 99.9%, the performance required in high-precision laser cavities.

Retroreflectors are devices that return light directly back toward its source regardless of the angle of arrival. Corner-cube retroreflectors, which use three mutually perpendicular mirror surfaces, are embedded in highway signs, bicycle reflectors, and the retroreflector arrays left on the lunar surface by Apollo astronauts. A laser beam fired at the lunar retroreflectors returns to within meters of the source telescope after a round trip of nearly 800,000 kilometers, enabling lunar laser ranging measurements of the Moon's distance to millimeter precision.

Refraction and Why Light Bends at Interfaces

When light crosses from one transparent material into another, it changes speed, and that change of speed produces a change of direction. This bending at an interface is called refraction, from the Latin for to break, and it is the fundamental mechanism behind every lens, every prism, and every optical fiber in the world.

To understand why a speed change causes a direction change, imagine a squad of soldiers marching in a straight line across a paved road. They approach a field of soft mud at an angle rather than head-on. The soldiers on the left reach the mud first and immediately slow down. The soldiers on the right are still on the pavement at full pace.

Because one end of the line is moving faster than the other, the entire line pivots: the fast end swings forward relative to the slow end. The squad is now marching in a different direction. This is exactly what happens to a wavefront of light when it crosses at an angle into a medium where it travels more slowly: one part of the wavefront slows before the rest, and the whole wavefront pivots toward the slower medium.

The refractive index of a material, usually written n, is the ratio of the speed of light in vacuum to the speed of light in that material. Vacuum has a refractive index of exactly 1. Air is very close to 1, at about 1.0003. Water has a refractive index of roughly 1.33, meaning light travels about 25 percent slower in water than in a vacuum. Common optical glass ranges from about 1.45 to 1.9, depending on composition. Diamond is famously around 2.4, which is why it bends light so dramatically and creates such intense sparkle when faceted. The higher the refractive index, the more a material slows light, and the more it bends a ray traveling into it at an angle.

This is why the pool makes its floor appear shallower than it really is: light traveling upward from the pool bottom bends away from the normal as it exits the water into air, and the eye traces those bent rays backward, assuming straight-line travel. The eye places the source at a shallower, horizontally shifted position. The fisherman's spear must be aimed at the real fish, not the image, because the image lies somewhere the fish is not.

Diagram 3.2 - Refraction at a Water-Air Interface

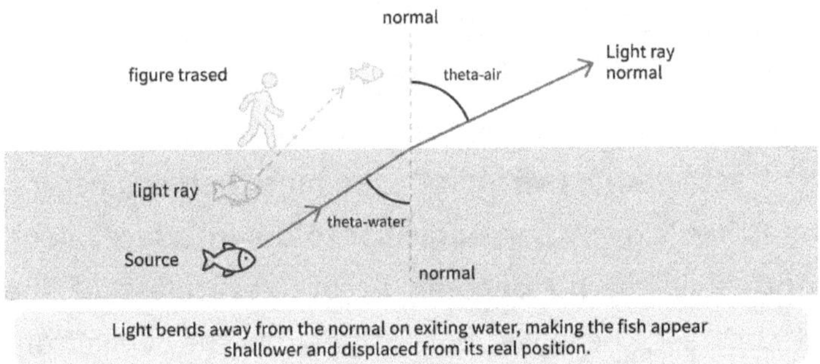

Temperature and pressure variations in the atmosphere produce spatially varying refractive indices, causing light to bend continuously as it passes through the atmosphere. This is atmospheric refraction. Stars near the horizon appear slightly higher in the sky than they geometrically are, because the atmosphere bends their light upward. Mirages on hot roads form when a layer of superheated air at pavement level has a lower refractive index than the cooler air above it, causing nearly horizontal rays from the sky to curve upward before reaching the observer's eye. The brain interprets this as a reflection from a water surface: a classic mirage.

Snell's Law in Plain Language

Refraction follows a precise, predictable rule known as Snell's law, named for the seventeenth-century Dutch mathematician Willebrord Snellius. Snell's law governs the extent of bending at an interface, based on the refractive indices of the two materials and the angle of incidence.

The conceptual core is completely intuitive without numbers. When light travels from a lower-index medium into a higher-index medium, from a faster medium into a slower one, it bends toward the normal. When it travels from a higher-index medium into a lower-index one, from slow to fast, it bends away from the normal. The greater the contrast in refractive indices, the more bending occurs. If light arrives exactly perpendicular to the interface, along the normal itself, no bending occurs regardless of the index contrast, because both sides of the wavefront change speed simultaneously.

This directional logic simultaneously explains several familiar phenomena. A glass of water makes a submerged spoon look broken at the waterline because the ray from the spoon exits into the air and bends away from the normal. A biconvex lens focuses light because its curved surfaces present the glass to parallel rays at angles that systematically bend them inward toward a common focal point. Starlight bends toward vertical as it descends through progressively denser atmospheric layers, each interface bending the ray a small amount toward the normal of that layer.

Diagram 3.3 - Snell's Law at a Glass-Air Interface

Refraction of light from air into glass

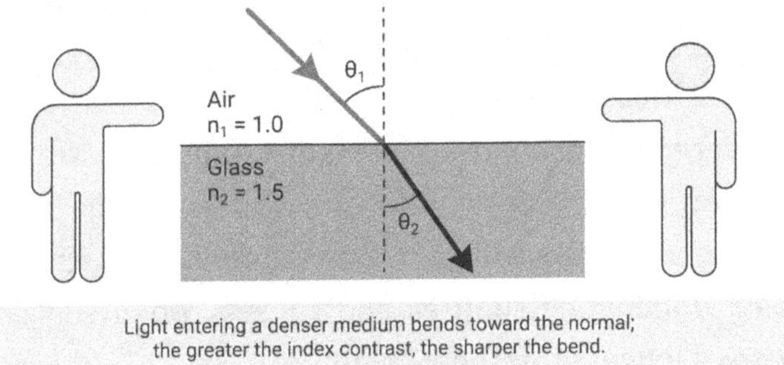

Light entering a denser medium bends toward the normal; the greater the index contrast, the sharper the bend.

Snell's law also governs dispersion, the variation of refractive index with wavelength. In most transparent materials, short-wavelength light, such as violet, experiences a slightly higher refractive index than long-wavelength light, such as red. Violet bends more than red at every interface. The Abbe number of a glass quantifies the magnitude of this variation across the visible spectrum. When a single lens refracts white light, violet and red are focused at different distances, causing color fringing around the edges of the image, known as chromatic aberration. Correcting chromatic aberration requires combining glass elements whose dispersive behaviors partially cancel each other, a technique called achromatic doublet design, and it is exactly the problem the lens designer in the opening scenario is solving.

Rayleigh scattering is related to but distinct from dispersion. When light encounters particles or density fluctuations much smaller than its wavelength, such as atmospheric gas molecules, it is scattered in all directions, with short wavelengths scattered far more strongly than long wavelengths. The intensity of Rayleigh scattering rises steeply with frequency, so blue and violet light scatter roughly ten times more than red under the same conditions. This is why the sky is blue: sunlight enters the atmosphere, and its short-wavelength components scatter in all directions, filling the sky with scattered blue light, while red and orange light travel more nearly straight through, dominating sunrises and sunsets when sunlight traverses a longer atmospheric path.

Diagram 3.4 - Mirage Formation on a Hot Road

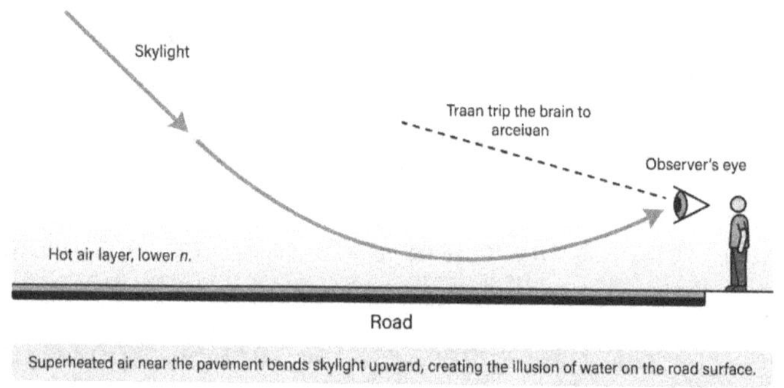

Superheated air near the pavement bends skylight upward, creating the illusion of water on the road surface.

Total Internal Reflection and the Fiber Optic Principle

Snell's law has a remarkable consequence that becomes apparent when light travels from a denser medium into a less dense one, and the angle of incidence is gradually increased from zero. At small angles, the refracted ray bends away from the normal as expected, and most light crosses the interface. As the angle increases, the refracted ray swings farther from the normal toward the horizon. At a specific angle called the critical angle, the refracted ray lies exactly along the interface itself. Beyond this angle, no refracted ray exists. Every bit of the light reflects into the denser medium. This is total internal reflection, and the word "total" is fully justified: there is no partial transmission, no leakage; 100% of the light is reflected. The critical angle depends on the ratio of the two refractive indices: the greater the contrast, the smaller the critical angle, and the more likely total internal reflection becomes.

The practical consequence of total internal reflection is optical fiber. An optical fiber is a thin strand of very pure glass consisting of a central cylindrical core surrounded by a layer called the cladding. The core has a slightly higher refractive index than the cladding. Light launched into the core at an angle shallow enough to exceed the critical angle at the core-cladding interface bounces off that interface with perfect reflectivity, then off the opposite wall, zigzagging along the fiber indefinitely without losing energy to refraction out through the sides. The fiber can be curved, bent, or wound onto a spool, and as long as the bends are not so sharp that the angle of incidence drops below the critical angle, the light stays trapped inside.

Diagram 3.5 - Total Internal Reflection in an Optical Fiber

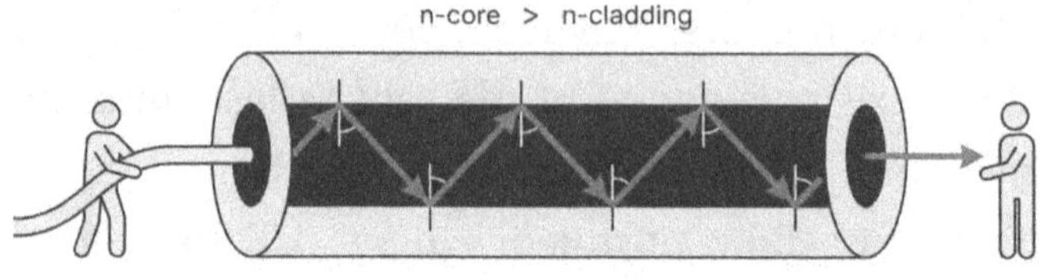

Light bounces along the fiber core without escaping because every reflection exceeds the critical angle.

The purity requirements for fiber optic glass are extraordinary. Conventional window glass attenuates light so severely that a pane only centimeters thick absorbs a noticeable fraction. Fiber-optic glass must be sufficiently pure for light to travel tens of kilometers before losing half its intensity. Modern single-mode silica fiber achieves attenuation of roughly 0.2 decibels per kilometer at the standard telecommunications wavelength of 1550 nanometers, allowing a signal to traverse about 15 kilometers before dropping to half power. In transoceanic cables, optical amplifiers called erbium-doped fiber amplifiers are placed at intervals of 60 to 100 kilometers to boost the signal and compensate for accumulated loss.

Evanescent waves are a subtle but important aspect of total internal reflection. When total internal reflection occurs, the light does not stop abruptly at the interface. An optical field penetrates only a very short distance into the lower-index medium as an exponentially decaying field, called an evanescent wave, from a Latin root meaning "to vanish". Under ideal conditions, it carries no net energy flow across the interface. However, if a second high-index medium is brought within the penetration depth of the evanescent field, on the order of hundreds of nanometers, the field can couple into it and transfer energy. This is called frustrated total internal reflection or optical tunneling. It is exploited in near-field scanning optical microscopy, which probes the evanescent field with a nanoscale tip to achieve spatial resolution far below the diffraction limit. It also appears in integrated photonic chips, where two closely spaced waveguides exchange power through their overlapping evanescent fields, enabling directional couplers, power splitters, and wavelength-selective routers.

Dispersion, Prisms, and Why Rainbows Form

When a narrow beam of white light enters one face of a triangular glass prism, it separates into a fan of colors as it exits the other face: violet at the most bent end, then blue, green, yellow, orange, and red at the least bent end. The ordering always matches the visible spectrum from short wavelength to long. The mechanism is dispersion, the dependence of a material's refractive index on the wavelength of light passing through it. In most transparent solids and liquids, short-wavelength light experiences a slightly higher refractive index than long-wavelength light, so violet is slowed and bent a little more than red at every surface. A prism has two tilted surfaces, and the bending at each accumulates, separating colors visibly over the distance involved.

Diagram 3.6 - Prism Dispersion Fan

Propension white Prism

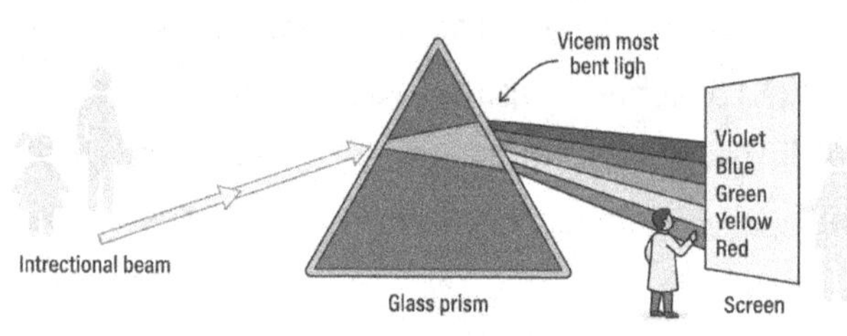

Dispersion separates white light because shorter wavelengths slow more in glass and bend more at each surface.

Rainbows are nature's prism demonstration, operating on a scale that fills a large arc of sky. Sunlight enters a spherical water droplet, refracts at the front surface with different wavelengths bending by different amounts, reflects off the back interior wall, and refracts again, exiting through the front. The two refractions and one internal reflection redirect the light back toward the sun at a wavelength-dependent angle: red exits at about 42 degrees from the incoming sunlight direction, violet at about 40 degrees. Every droplet in the sky performs this separation simultaneously, each sending red from one direction and violet from another, producing a colored arch with red on the outside and violet on the inside.

Thin-film interference produces related but mechanically distinct color effects. When light hits a thin, transparent film, part of it reflects from the top surface and part from the bottom. If the path-length difference between the two reflected beams is comparable to the wavelength, they interfere constructively for some wavelengths and destructively for others. This selective amplification and cancellation is what colors an oil slick. A film one-half wavelength thick for green light cancels green in reflection and appears magenta. Tilting your head changes the effective film thickness you are viewing through, which is why the colors shift as you move. Anti-reflection coatings are designed using the same principle, with precisely controlled film thicknesses to ensure destructive interference for the wavelengths the designer wants to suppress.

Diagram 3.7 - Thin-Film Interference on a Parking-Lot Oil Slick

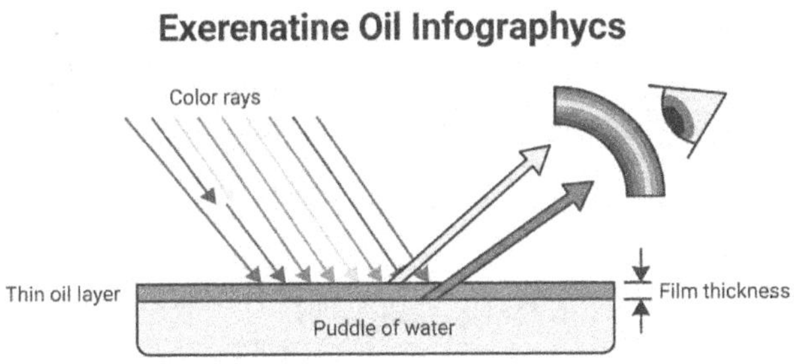

Exerenatine Oil Infographycs

Color rays

Thin oil layer

Film thickness

Puddle of water

Reflections from the top and bottom of a thin oil film interfere, canceling some wavelengths and amplifying others to create vivid color.

Diffraction Gratings and Spectroscopy

A diffraction grating is a surface ruled with thousands of parallel lines per millimeter. When light strikes it, each line acts as a source of scattered waves that interfere with waves from neighboring lines. For most angles and wavelengths, these waves cancel. But at specific angles that depend on wavelength and line spacing, the waves from all lines arrive exactly in phase and reinforce each other strongly. These reinforced directions are called diffraction orders, and different wavelengths produce their peak reinforcement at different angles, separating a mixed-color beam into its constituent wavelengths with a precision far beyond what a prism can achieve.

The physical mechanism underlying a diffraction grating is Huygens' principle, formulated by the seventeenth-century Dutch physicist Christiaan Huygens. Huygens proposed that every point on a wavefront can be treated as a new source of secondary spherical waves, and the subsequent wavefront is the surface tangent to all those secondary waves. When a wavefront passes through a narrow opening or across the edge of a surface, the secondary waves spread in all directions, not just the original direction of travel. This spreading around corners and edges is diffraction, and Huygens' principle correctly predicts both the directions of maximum intensity and cancellation in any diffraction pattern.

Diagram 3.8 - Diffraction Grating Separating Colors

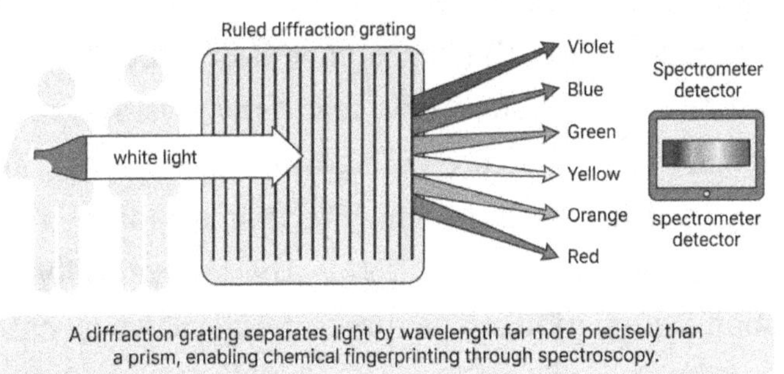

A diffraction grating separates light by wavelength far more precisely than a prism, enabling chemical fingerprinting through spectroscopy.

Spectrometers built around diffraction gratings are among the most broadly used analytical instruments in science and industry. By precisely measuring the angles at which specific wavelengths emerge from the grating, a spectrometer determines which wavelengths are present in the incoming light and their intensities. This spectral fingerprint is immensely informative because every element and compound absorbs and emits light at characteristic wavelengths. Astronomers use spectrometers attached to telescopes to determine the chemical composition of stars, the velocity of galaxies, and the atmospheric composition of exoplanets. Chemists use them to identify compounds. Medical diagnosticians use them to analyze blood and tissue samples. Environmental monitors use them to identify pollutants in air and water by the wavelengths at which specific molecules absorb.

Diffraction from two-dimensional periodic structures extends this principle into three dimensions. The regular array of atoms in a crystal acts as a three-dimensional diffraction grating for X-rays, whose wavelengths are comparable to the atomic spacing. By measuring the angles and intensities of X-rays diffracted from a crystal, scientists reconstruct the three-dimensional positions of every atom in the crystal's unit cell. X-ray crystallography determined the structure of proteins, vitamins, and penicillin, and it was the technique that established the double-helix structure of DNA. The universality of diffraction from periodic structures, at optical wavelengths for ruled gratings and at X-ray wavelengths for crystals, is one of the most elegant demonstrations of how the single-wave principle scales across radically different length scales.

Interference: Waves Adding and Canceling

When two or more waves of the same frequency overlap in the same region of space, their amplitudes add together at every point. Where two wave crests coincide, the combined wave is brighter than either alone: this is constructive interference. Where a crest from one wave coincides with a trough from another, the amplitudes cancel: this is destructive interference. The result is a pattern of alternating bright and dark regions called interference fringes, whose spacing and orientation encode information about the waves that produced them. Interference is the behavior that most distinctly marks light as a wave rather than a stream of classical particles.

The classic demonstration is the double-slit experiment, first performed by the English scientist Thomas Young in 1801. Young allowed a beam of light to pass through a card with two narrow parallel slits and observed the pattern on a screen behind the card. Rather than seeing two bright stripes, he saw a series of alternating bright and dark bands spanning a width much larger than the two-slit gap. Bright bands formed where light from the two slits arrived with path lengths differing by a whole number of wavelengths, so crests aligned and waves added. Dark bands formed where path lengths differed by a half-integer number of wavelengths, so crests met troughs and canceled. Young used the fringe spacing and the geometry of his apparatus to calculate the wavelength of light, obtaining a value close to modern measurements. This was the first direct measurement of the wavelength of light, and simultaneously the first definitive demonstration that light is a wave.

Diagram 3.9 - Young's Double-Slit Interference Pattern

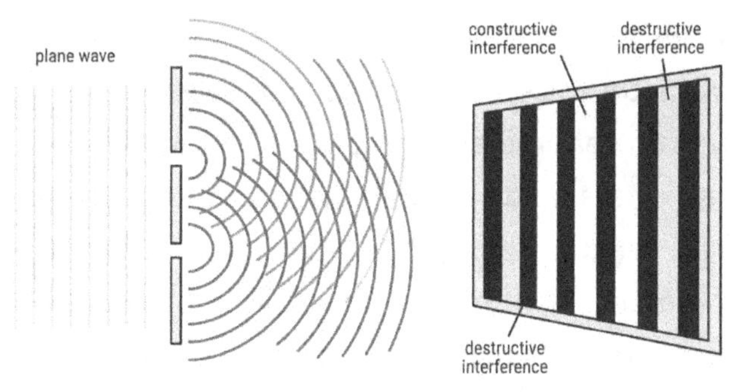

Two slits produce overlapping wavefronts that create a striped fringe pattern, proving light behaves as a wave.

Coherence is the property that enables stable interference. Two light waves interfere stably only if they maintain a fixed phase relationship over time and across the beam. A light source is coherent if the phase of its output is predictable over a sufficient time interval and spatial extent. Lasers are highly coherent: they emit light in which all the photons are in phase across the beam, called spatial coherence, and they emit a very narrow range of frequencies, called temporal coherence, so the wave train stays self-consistent over long distances. Sunlight and incandescent bulbs emit light with very short coherence lengths, meaning the phase relationship between different points in the beam breaks down within a fraction of a micrometer. This is why it is difficult to make white light from a conventional source interfere with itself unless the path-length difference between the two arms of an interferometer is nearly zero.

Coherence length is the maximum path-length difference over which stable interference can still be observed. For sunlight, coherence length is only a few micrometers, which is why oil-slick colors appear only in films a fraction of a micrometer thick. For a typical diode laser, it might be millimeters to centimeters. For ultra-stabilized lasers used in optical clocks and gravitational-wave detectors, the coherence length can reach millions of kilometers. Interferometers, instruments that split a beam into two paths and then recombine them to observe the interference pattern, use this sensitivity to measure path-length differences at the nanometer and sub-nanometer level. The LIGO gravitational-wave detectors are interferometers with four-kilometer arms that detect a spacetime stretch smaller than one-thousandth of a proton diameter by observing interference fringes shift by a corresponding fraction of a fringe.

The Diffraction Limit and the Resolution of Optical Instruments

Diffraction imposes a fundamental physical limit on the sharpness of any image formed by a lens or mirror. This is the diffraction limit, and it is not an engineering shortcoming that better manufacturing could overcome. It is a consequence of the wave nature of light and applies to every optical system, regardless of how perfect the optics are. Understanding the diffraction limit matters not just for telescope and microscope designers, but for anyone working with photonic sensors, optical communications equipment, or quantum imaging systems, because it defines the ultimate boundary of what classical optics can achieve.

When light passes through a circular aperture such as a lens, diffraction causes the wavefront to spread slightly at the aperture edge. The light from all points around the aperture rim diffracts inward, and when these contributions are added up, the result at the focal point is not a geometric point but a small bright disk surrounded by concentric bright and dark rings. This pattern is called an Airy disk, named for the nineteenth-century British Astronomer Royal George Biddell Airy, who calculated its form. The radius of the central disk depends on two quantities: the wavelength of the light and the diameter of the aperture. Shorter wavelengths and larger apertures both produce smaller Airy disks, meaning finer detail in the image.

Diagram 3.10 - Airy Disk Diffraction Pattern

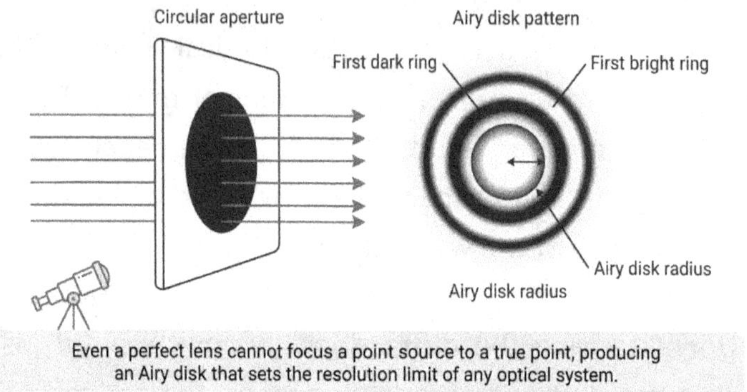

Even a perfect lens cannot focus a point source to a true point, producing an Airy disk that sets the resolution limit of any optical system.

The Rayleigh criterion is the conventional standard for when two nearby point sources are just barely resolved. Two sources are considered resolved when the centers of their Airy disks coincide with each other's first dark rings. If the two disks overlap more closely, the combined image appears as a single blurred blob, and the sources cannot be distinguished. This criterion sets a hard lower limit on the angular resolution of a telescope: a larger mirror produces smaller Airy disks and resolves finer angular detail. It is why ground-based observatories and space telescopes are built with the largest mirrors funding and engineering allow, and why the James Webb Space Telescope's 6.5-meter primary mirror resolves details in distant galaxies that were invisible to the Hubble Space Telescope's 2.4-meter mirror.

In conventional optical microscopy, the smallest resolvable feature is approximately half the wavelength of the illuminating light. For green light at 550 nanometers, this is about 275 nanometers. Many biologically and technologically important structures are smaller: individual proteins, viruses, and the features of modern semiconductor circuits are all well below the diffraction limit for visible light. This barrier stood for over a century as the fundamental limit of optical imaging until a series of techniques developed from the 1990s onward, now grouped as super-resolution microscopy, circumvented it by exploiting the nonlinear optical properties of fluorescent molecules. These techniques earned the 2014 Nobel Prize in Chemistry and are discussed in the chapter on quantum sensing and imaging.

In semiconductor lithography, the feature size that can be printed on a silicon wafer is set by the diffraction limit of the exposure optics. The entire history of semiconductor scaling is largely a history of shrinking the illumination wavelength: from near-ultraviolet to deep ultraviolet to extreme ultraviolet at 13.5 nanometers, each step allowing finer features to be printed. Understanding the physical reason the limit exists, which is the wave nature of light and the spreading of wavefronts at aperture edges, is what allows engineers to reason correctly about which circumvention strategies are physically sound.

Diagram 3.11 - Telescope Diffraction Limit: Small vs. Large Aperture

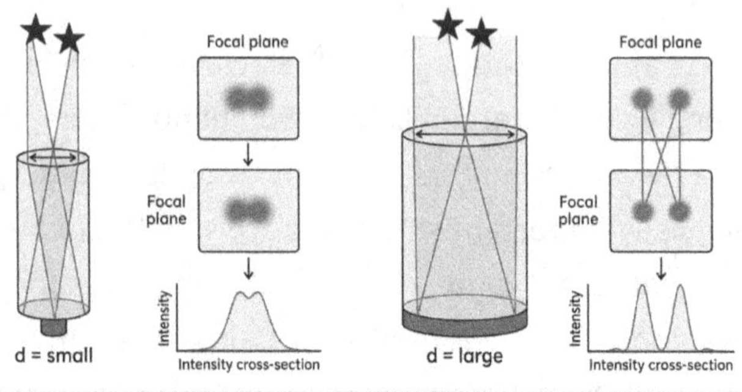

Larger apertures produce smaller Airy disks and finer resolution, explaining why every telescope mirror size increase improves image sharpness.

What This Means For You

If you work in optical engineering, fiber communications, or photonic device design, the concepts in this chapter are a daily working vocabulary. Every time you specify the numerical aperture of a fiber or lens, you are expressing an angular acceptance cone determined by total internal reflection or geometric optics. Every time you evaluate a spectrometer, you are asking about the grating groove density and its consequences for spectral resolution. Every time you review an anti-reflection coating specification or a laser linewidth number, you are reading the consequences of thin-film interference and coherence length. This chapter has given those numbers conceptual homes.

For engineers in semiconductor manufacturing, the diffraction limit is perhaps the most important single constraint in the field. The full history of Moore's Law scaling is largely a history of pushing against this limit by shrinking wavelength, increasing numerical aperture, and developing resolution-enhancement techniques. Understanding the physical reason the limit exists allows engineers to determine which circumvention strategies are physically sound and which are dead ends.

For those working at the intersection of photonics and biology, near-field imaging, evanescent-wave microscopy, and structured-illumination techniques all exploit the behaviors of light at interfaces and apertures described in this chapter. The ability to see individual proteins in a living cell or to track viral particles in real time depends entirely on manipulating the same wave and interference properties introduced here.

For quantum optics practitioners, this chapter provides the classical reference frame against which quantum behavior becomes meaningful. Quantum interference, which appears in the behavior of single photons at beam splitters and in quantum state measurement, is the quantum mechanical version of the wave interference described here. In classical wave optics, interference is a property of the amplitude of an electromagnetic field. In quantum optics, interference occurs even for single photons, one at a time, reflecting the probabilistic nature of quantum measurement rather than the addition of classical field amplitudes. You cannot understand that distinction without first understanding what classical interference is and what it produces.

Diagram 3.12 - Evanescent Wave Coupling in a Photonic Waveguide

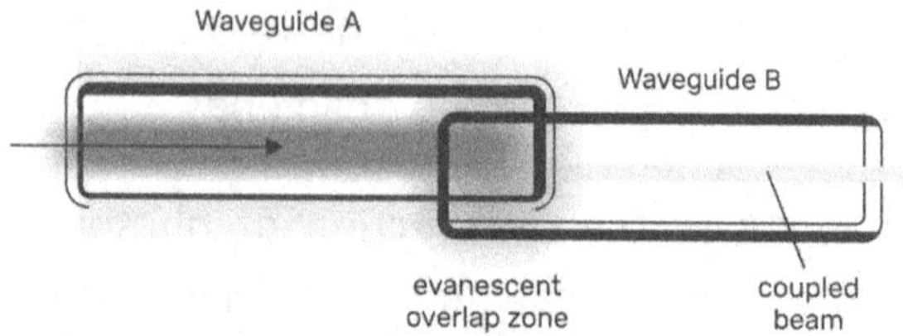

Evanescent fields extend beyond waveguide walls, allowing nearby waveguides to exchange light energy without physical contact.

Takeaway

Light does not travel blindly through the world. At every boundary between materials and at every aperture, it responds to its environment according to precise and predictable rules. Reflection preserves the angle of a light beam at a smooth surface, making images possible and laser cavities functional. Refraction bends light at interfaces in proportion to the change in speed, governed by Snell's law, enabling lenses to focus, fibers to guide, and the atmosphere to bend starlight. Total internal reflection is the extreme case, turning an interface into a perfect mirror when the angle of incidence exceeds the critical angle, making transoceanic optical fiber communication possible. Dispersion separates wavelengths in every refractive process, producing rainbows, chromatic aberration, and the spectral signatures that spectrometers decode. Diffraction causes light to spread at apertures and edges, governed by Huygens' principle, enabling gratings to separate wavelengths with exquisite precision while simultaneously imposing the diffraction limit on the resolution of every optical instrument ever built. Interference occurs when multiple wavefronts overlap, with constructive and destructive outcomes that encode path-length differences down to fractions of a nanometer.

These behaviors are not a list to memorize. They are a unified picture of how light interacts with the physical world, each one flowing from the same fundamental fact: light is an electromagnetic wave with amplitude, frequency, phase, and wavelength that govern how it propagates, bends, combines, and spreads. The same wave nature that gives rise to interference fringes in Young's two-slit experiment also gives rise to the quantum interference of single photons at a beam splitter in a quantum computing circuit. The same Huygens principle that explains diffraction gratings underlies the angular resolution of quantum imaging systems. The classical and quantum behaviors are not separate realms: they are the same physics viewed at different scales and intensities.

The lens designer correcting chromatic aberration, the fiber engineer specifying a splice angle, the astronomer calculating telescope resolution for a binary star, the lithographer selecting an illumination wavelength: all are applying what this chapter has laid out. All are working with light as a wave that refracts, reflects, diffracts, and interferes according to the geometry and composition of whatever it encounters. That picture is now yours. The next chapter turns from classical wave optics to the instrument that exploits coherence and stimulated emission to produce light of a quality that classical descriptions can only characterize from the outside: the laser.

4 Lasers: The Most Precise Light on Earth

Opening Scenario

The room smells of antiseptic and cool, circulated air. You are lying in a reclining chair, a soft brace holding your head still, a foam ring keeping your eyelids gently open. A blinking orange fixation light is the only feature on the ceiling above you. The surgeon tells you to relax. She explains that a laser is about to reshape the surface of your cornea, the transparent dome at the very front of your eye, to correct the curvature that has been making distant objects blurry your entire life. Twenty seconds per eye. You nod, but the word laser sits in your mind with an unfamiliar weight. You have used laser pointers. You know lasers cut steel in factories. Now one is aimed at your eye.

The surgeon is wielding a femtosecond laser, one of the most precise tools ever created. Each pulse lasts less than one quadrillionth of a second. A femtosecond is to a second what a second is to roughly thirty-two million years. The pulse deposits its energy into the corneal tissue and vanishes before surrounding cells have time to register heat: no burning, no charring, no collateral damage. The laser severs molecular bonds in a layer thinner than a human hair and leaves every adjacent cell completely untouched. You will leave the clinic in two hours, seeing better than you have since childhood. What made that possible is the subject of this chapter.

Diagram 4.1 - LASIK Eye Surgery Scene

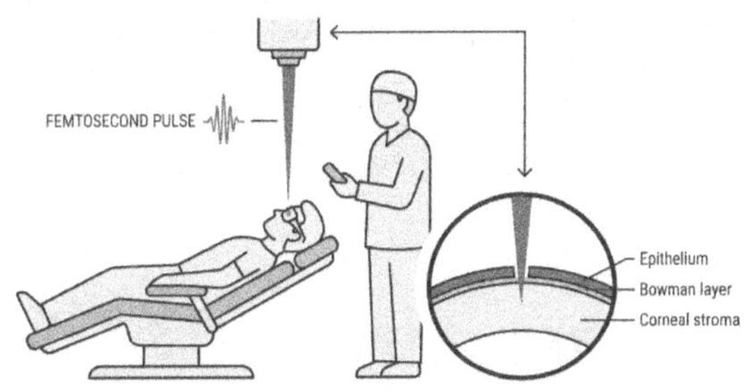

A femtosecond laser reshapes the cornea with sub-micron precision, correcting vision in under twenty seconds per eye.

Why It Matters

The laser is arguably the most consequential photonic invention of the twentieth century, yet most people interact with lasers dozens of times each day without knowing it. Every barcode scan, every fiber-delivered video stream, every laser-printed page depends on one. The technology has woven itself so thoroughly into manufacturing, medicine, communications, and research that removing it would be roughly as disruptive as removing electricity. And every laser, from a five-dollar pointer to a billion-dollar industrial cutting machine, runs on the same quantum-mechanical process that Albert

Einstein predicted in 1917, decades before anyone built a device to exploit it.

Understanding lasers is the conceptual gateway to everything that follows in this book. Lasers are the primary light sources in fiber optic communications, quantum key distribution, gravitational-wave detection, photonic quantum computing, and the optical atomic clocks that define our most precise measurement of time. To understand any of those technologies, you first need to understand what makes laser light different from every other kind of light, and why that difference matters so profoundly.

What Makes Laser Light Different: Coherence, Monochromaticity, and Directionality

Ordinary light is a wild mixture of photons spanning many wavelengths, radiating in all directions, with wave crests at completely arbitrary positions bearing no relationship to one another. A laser produces light radically different in all three of those respects, and those differences give it the precision to reshape a cornea or carry terabits of data across an ocean.

The first difference is called monochromaticity, which means one color. Laser light contains photons of a single wavelength, or at most an extremely narrow range. A red laser pointer, for example, might emit at 650 nanometers with a spread of less than one nanometer in either direction. Compare that to a red LED, which emits across a band twenty to forty nanometers wide. The laser's extreme color purity is what allows it to carry precise frequency information, measure distances to the nearest nanometer, and interact with specific atomic transitions without disturbing neighboring ones.

The second difference is coherence, the most important property for applications throughout this book. Temporal coherence means that the wave peaks and troughs maintain a fixed phase relationship over time: photons march in step. The distance over which this orderly pattern persists is the coherence length. A sodium street lamp has a coherence length of a few millimeters; a high-quality single-frequency laser can have a coherence length of hundreds of kilometers. Spatial coherence means the beam's wavefront is smoothly ordered, allowing the beam to be focused to a tiny spot. This is why a laser can burn a hole in steel while a flashlight of the same total power would barely warm your hand.

The third difference is directionality. A laser beam propagates along a single axis with an angular spread so small that it can travel kilometers and remain the same diameter, a direct consequence of spatial coherence and the optical cavity geometry. This allows coupling into the microscopic core of an optical fiber with almost no loss.

Diagram 4.2 - Coherence Comparison: Ordinary Light vs. Laser Light

Incoherent | **Laser apem**

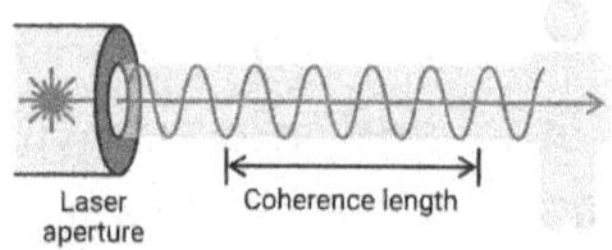

Incoherent: many wavelengths, random phases, wide angular spread.

Coherent: single wavelength, fixed phase, near-zero angular spread.

Laser light is monochromatic, coherent, and directional; ordinary light is none of these.

Einstein's 1917 Prediction: Stimulated Emission

The story of the laser begins not in 1960 but in 1917, when Albert Einstein published a theoretical analysis of how atoms interact with light. Einstein identified three distinct processes. The first is absorption: a photon with exactly the right energy passes near an atom in its ground state; the atom absorbs the photon and jumps to a higher energy level, called an excited state, storing energy like a compressed spring.

The second process is spontaneous emission. After a brief time in an excited state, an atom randomly releases a photon and drops back to its ground state. The word spontaneous is key: the photon flies off in a random direction with a random phase, bearing no coordinated relationship to any other photon. This is what ordinary light bulbs, flames, and stars do. The result is incoherent light.

The third process is the one Einstein added, and it is the operating principle of every laser ever built. He called it stimulated emission. When a photon passes near an atom already in an excited state, and the photon's energy matches the gap between the excited and lower states, something remarkable happens: instead of being absorbed, the photon triggers the atom to release its stored energy as a second photon that is an exact copy of the first. Same wavelength, same direction, same phase. Two photons now travel together where one entered. Those two can trigger two more excited atoms, producing four. Four becomes eight. This is optical amplification, the heart of every laser. The name "laser" is an acronym for the mechanism "Light Amplification by Stimulated Emission of Radiation."

Einstein's insight was purely theoretical in 1917: no one knew how to exploit it. That would take four more decades. But the mechanism he described is exactly what happens inside every laser you have ever encountered.

Diagram 4.3 - Einstein's A and B Coefficients: Three Light-Matter Processes

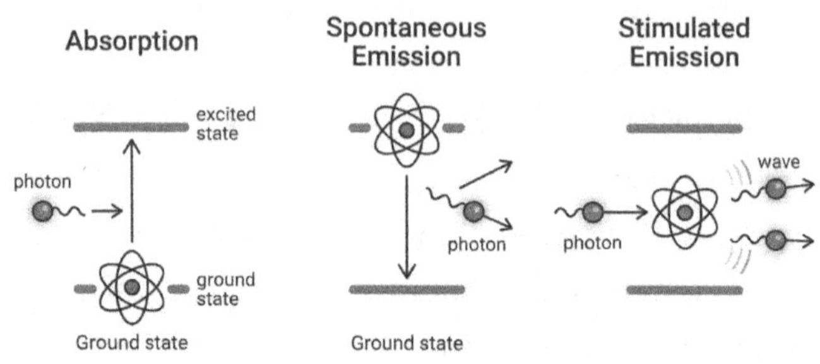

Stimulated emission is the key: one photon in, two identical photons out.

Population Inversion: The Counterintuitive Prerequisite

There is a critical obstacle between Einstein's prediction and a working laser: thermodynamics. Under normal conditions, most atoms sit in their ground state. A passing photon is far more likely to be absorbed than to trigger stimulated emission. To make stimulated emission dominate, the engineer must create a population inversion: more atoms in the excited state than in the ground state. Achieving this is the central engineering challenge of laser design.

A population inversion cannot occur naturally in a two-level system because equilibrium prevents the excited population from ever exceeding the ground-state population. Real lasers circumvent this using three-level or four-level energy schemes. In a three-level laser, atoms are pumped to a high-energy short-lived state and rapidly decay to a longer-lived metastable level. Because this level has a relatively long lifetime, atoms accumulate there. If enough energy is pumped fast enough, more atoms pile up in the metastable level than remain in the ground state: population inversion is achieved.

Four-level schemes are even more efficient. The lower laser transition level is not the ground state but a fourth level just above it. Atoms drop into this fourth level after stimulated emission, then rapidly decay non-radiatively to the ground state, keeping the lower level nearly empty and making population inversion easy to maintain. Most modern solid-state and fiber lasers use four-level or quasi-four-level schemes for exactly this reason.

The energy needed to create population inversion is supplied by pumping. Gas lasers use an electrical discharge. Solid-state lasers are pumped optically by flashlamps or diode lasers. Semiconductor diode lasers are pumped electrically by current through the junction. The method varies, but the goal is always the same: get more atoms excited than resting.

Diagram 4.4 - Population Inversion Bar Chart: Excited vs. Ground State

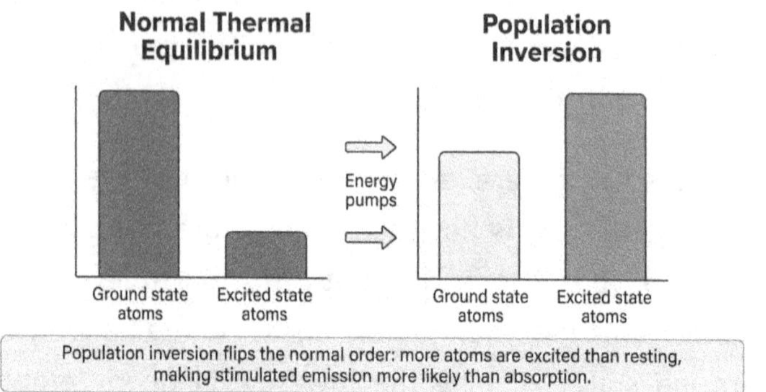

Population inversion flips the normal order: more atoms are excited than resting, making stimulated emission more likely than absorption.

The Optical Cavity: Selecting One Wavelength and Building the Beam

Stimulated emission alone would produce amplified light in all directions: bright but incoherent. To get a laser beam, you need something that selects a single direction and wavelength, repeatedly forces light through the gain medium, and allows a controlled fraction of the light to escape. That is the optical cavity, also called the optical resonator.

In its simplest form, an optical cavity consists of two mirrors placed at opposite ends of the gain medium and aligned with extreme precision. When a photon produced by stimulated emission travels along the cavity axis, it bounces off one mirror, passes back through the gain medium picking up more stimulated-emission copies of itself, bounces off the other mirror, and repeats. With each pass, the number of photons grows. Photons traveling in other directions quickly escape through the sides and are lost. Only the axially traveling photons survive and multiply.

The cavity also imposes strict wavelength selection through longitudinal modes: the specific wavelengths that form stable standing wave patterns inside the cavity. Like a guitar string that vibrates only at wavelengths fitting a whole number of half-cycles between its fixed ends, an optical cavity allows only wavelengths that fit an integer number of half-cycles between its mirrors to build up. All other wavelengths cancel destructively and are extinguished. By carefully choosing cavity length and mirror coatings, laser designers restrict the output to a single mode and a near-perfectly monochromatic beam.

One mirror in each cavity is partially transmissive, reflecting most of the light into the gain medium while allowing a small percentage to pass through. This is the output coupler. Too much transmission and light escapes before it is fully amplified; too little, and the laser wastes energy. The gain medium, two mirrors, and the pumping mechanism are the three essential components of every laser, regardless of how the surrounding engineering varies.

Diagram 4.5 - The Optical Cavity: Mirrors, Gain Medium, and Output Coupler

Laser infographic icotanicotion

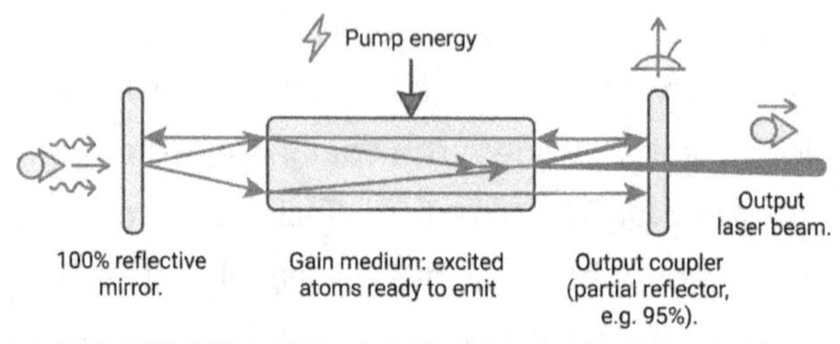

The cavity selects one wavelength and forces light through the gain medium repeatedly until it exits as a focused, coherent beam.

Types of Lasers: A Family Portrait

The word laser covers an astonishing range of devices, from a semiconductor chip smaller than a grain of rice consuming a few milliwatts to the fusion-research lasers that unleash tens of millions of joules in a single pulse. What unifies them all is stimulated emission in a population-inverted gain medium inside an optical cavity. What distinguishes them is everything else: the gain medium, the pumping method, the cavity design, the wavelength, the power level, and the intended application.

Gas lasers use a gas or a gas mixture as the gain medium. The helium-neon laser fills a glass tube with helium and neon and passes an electrical discharge through it. The discharge excites helium atoms, which transfer energy to neon atoms through collisions, creating a population inversion in the neon and producing a stable red beam. Carbon dioxide lasers use CO2 molecules as the gain medium and emit infrared light invisible to the human eye;

they can reach kilowatt power levels, making them effective for cutting and welding metals and ceramics.

Solid-state lasers use a transparent crystal or glass material doped with active ions. The most widely known is the neodymium-doped yttrium aluminum garnet laser, the Nd: YAG, whose neodymium ions produce a coherent beam at 1064 nanometers in the near-infrared when optically pumped. Nd: YAG lasers can be pulsed to extremely high peak powers, making them useful for rangefinding, materials drilling, and non-invasive medical procedures.

Semiconductor diode lasers are the most numerous type of laser on Earth. They are the light source in every optical disc drive, laser printer, barcode scanner, and fiber-optic transmitter, and they power most modern solid-state and fiber lasers. Their small size, high efficiency, and low cost have made them the dominant laser platform of the modern era.

Fiber lasers guide and amplify light within a doped optical fiber whose core serves simultaneously as the waveguide, the gain medium, and part of the optical cavity. Industrial fiber lasers now routinely deliver multiple kilowatts of continuous power in beams of exceptional quality, cutting stainless steel with edge precision that no mechanical tool can match.

Diagram 4.6 - Laser Family Tree: Gas, Solid-State, Semiconductor, and Fiber

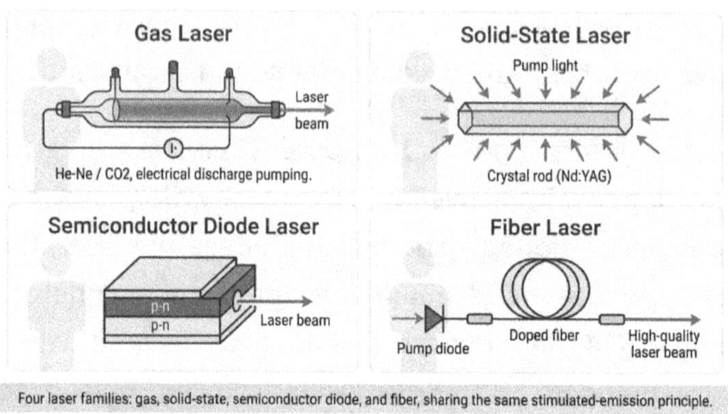

Four laser families: gas, solid-state, semiconductor diode, and fiber, sharing the same stimulated-emission principle.

The Semiconductor Diode Laser: A Billion Lasers in Your Pocket

Of all laser families, none has had a greater impact on daily life than the semiconductor diode laser. In the quantum description of a solid, electrons occupy energy bands. The valence band holds electrons bound to atoms; the conduction band holds electrons free to carry current. Between them sits the band gap, a range of energies electrons cannot have. A diode laser joins an n-type semiconductor layer (extra electrons) to a p-type layer (holes, or electron deficiencies). When a current is applied, electrons and holes are pushed toward the junction, where electrons fall from the conduction band into the valence band and emit photons with a wavelength set by the band gap.

In a light-emitting diode, this process produces spontaneous emission: random, incoherent photons in all directions. To make a diode laser, you need to add population inversion and an optical cavity. Population inversion is achieved by driving sufficient current through the junction to exceed a critical electron density threshold. The optical cavity is formed by the cleaved end facets of the semiconductor chip itself: the semiconductor's high refractive index creates a partial mirror at the air-semiconductor interface, reflecting a fraction of the light into the device.

The practical advantages of semiconductor diode lasers are staggering. A typical communication diode laser chip is roughly three hundred micrometers long. Modern designs convert more than seventy percent of electrical input into light, compared to a few percent for a traditional incandescent bulb. Diode lasers used in fiber optic systems switch on and off at rates of ten billion times per second or more, encoding digital data at many gigabits per second in the pulses of light traveling through the fiber. And they are inexpensive: mass manufacturing has driven the cost of a standard communication diode laser to a few dollars or less.

Diagram 4.7 - Semiconductor Diode Laser Cross-Section

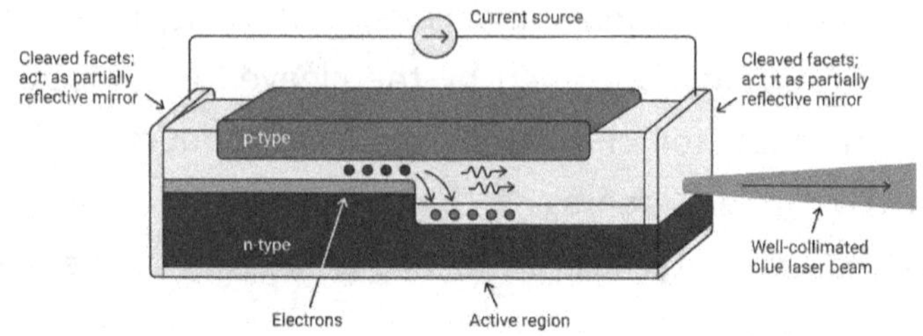

A diode laser uses cleaved semiconductor facets as its cavity mirrors; current drives electrons and holes to recombine and emit coherent photons.

Ultrafast Lasers: Stopping Time with Light

Conventional lasers produce continuous beams or pulses lasting microseconds or nanoseconds. In the world of atomic and molecular dynamics, even a nanosecond is an eternity. Chemical bonds vibrate at picosecond timescales. Electrons reorganize within atoms at femtosecond timescales, meaning quadrillionths of a second. To observe these processes in real time, you need a flash of light shorter than the event itself.

The technique that makes ultrashort pulses possible is called mode-locking. A cavity supports many longitudinal modes simultaneously; in a normal laser, these modes oscillate with random phases. Mode-locking forces all modes into phase at the same moment, producing a single intense peak. A moment later, the phases diverge, the peak disperses, and the cycle repeats: a regular train of extremely short pulses, each separated by the round-trip time of light inside the cavity.

Femtosecond pulses carry peak powers millions of times higher than the beam's average because all the energy is compressed into a brief instant and delivered before heat diffuses outward. Ultrafast spectroscopy uses a pump pulse to trigger a chemical reaction and a probe pulse arriving femtoseconds later to photograph the molecular geometry, producing a slow-motion movie of bond breaking and forming. Ahmed Zewail won the Nobel Prize in Chemistry in 1999 for pioneering this technique.

Attosecond pulses, each a thousandth of a femtosecond, have more recently made it possible to photograph the motion of electrons inside atoms for the first time.

Diagram 4.8 - Femtosecond Pulse Train from a Mode-Locked Laser

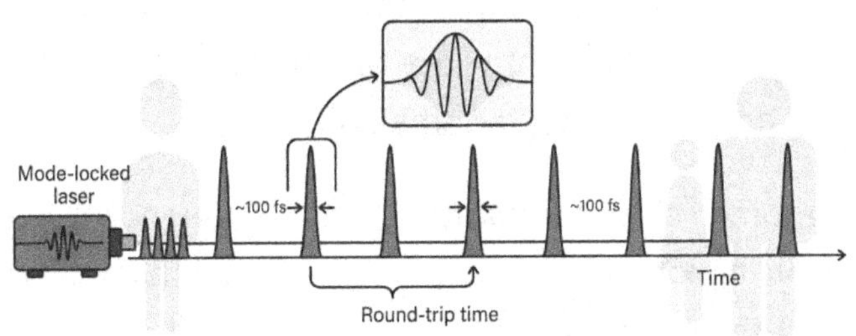

Mode-locking produces a regular train of ultrashort pulses; each pulse contains only a few optical cycles and lasts a fraction of a trillionth of a second.

The Frequency Comb: A Ruler Made of Light

One of the most surprising gifts of the mode-locked laser is the frequency comb. Analyze a mode-locked pulse train in the frequency domain, and you find that it consists of a large number of discrete frequency components, each separated from the next by the same gap. Plotted as intensity versus frequency, they look like the teeth of a comb: evenly spaced vertical lines across a broad range of optical frequencies.

Because the spacing between the teeth is determined solely by the repetition rate of the laser, which can be measured and controlled with extreme precision, the frequency comb provides a direct, calibrated link between the easily measured frequencies of radio waves and the much harder-to-measure optical frequencies of light. Before the frequency comb, measuring the precise frequency of a visible light wave required an enormously complex chain of frequency-doubling and mixing stages that filled an entire laboratory. With a frequency comb, you can measure an optical frequency in minutes by finding which comb tooth the unknown frequency is nearest to and measuring the small offset.

Optical atomic clocks calibrated using frequency combs would gain or lose less than one second in thirty billion years, and they can measure the gravitational redshift of general relativity across a height difference of just a few centimeters. John Hall and Theodor Hansch shared the Nobel Prize in Physics in 2005 for developing the optical frequency comb.

Diagram 4.9 - The Optical Frequency Comb

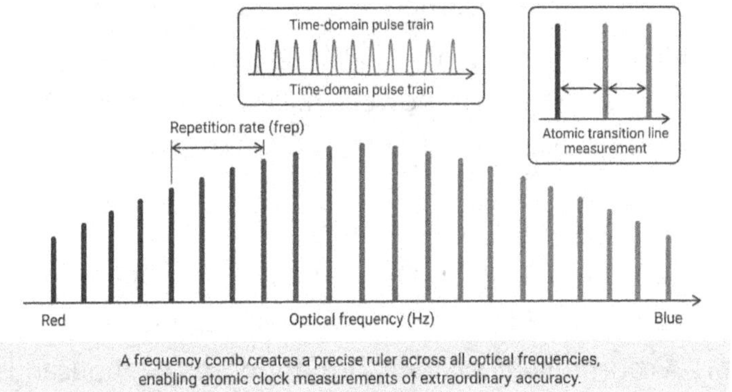

A frequency comb creates a precise ruler across all optical frequencies, enabling atomic clock measurements of extraordinary accuracy.

Controlling the Pulse: Q-Switching and Stored Energy

Mode-locking is not the only way to generate short, intense laser pulses. Another widely used technique is Q-switching, which produces pulses by deliberately preventing lasing for a period, storing energy in the gain medium, and then releasing it all at once in a single giant pulse. The Q in Q-switching stands for quality factor, a measure of how efficiently an optical cavity stores energy. A high-Q cavity stores energy with little loss per round trip; a low-Q cavity leaks energy rapidly.

In a Q-switched laser, the cavity is initially set to a low-Q state by inserting an element that blocks the intracavity light, preventing lasing even while the gain medium is being pumped. Energy accumulates in the excited atoms, building the population inversion far beyond the normal lasing threshold. Then, at a precisely chosen moment, the blocking element is switched open, typically by rotating a polarizing crystal, triggering an acousto-optic modulator, or using a saturable absorber. The stored energy erupts as a single intense pulse lasting nanoseconds to microseconds, with peak powers millions of times higher than the continuous output of the same laser.

Q-switched lasers are the workhorses of industrial laser processing, lidar rangefinding, and surgical applications that benefit from high peak power with modest average power. Airborne lidar systems fire hundreds of thousands of Q-switched pulses per second, measuring terrain topography by recording the round-trip time of each returned echo. The detailed elevation models that geologists, urban planners, and autonomous-vehicle systems rely on are built almost entirely by Q-switched lasers mapping the world from above.

Diagram 4.10 - Q-Switching: Storing and Releasing Energy

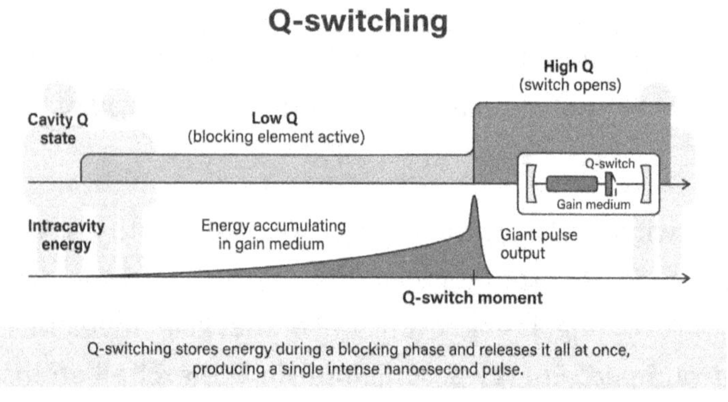

Q-switching stores energy during a blocking phase and releases it all at once, producing a single intense nanoesecond pulse.

Lasers Across Industries: A Technology That Transformed Civilization

Walk into any modern metalworking facility, and you will see a fiber or CO2 laser cutting machine directing a focused beam, often a kilowatt or more, along a computer-controlled path. At the same time, an assist gas blows molten metal away. The result is cut quality that mechanical saws and punches cannot match: smooth edges, tight tolerances, intricate geometries, and no tool-to-workpiece contact.

In medicine, lasers span the breadth of clinical practice. Argon laser photocoagulation seals leaking retinal blood vessels. Nd: YAG lasers clear secondary cataracts. Q-switched lasers remove tattoos by fragmenting ink pigment into sizes the immune system can clear. Erbium lasers ablate tooth enamel with less vibration than a drill. Photodynamic therapy combines a light-sensitive drug with targeted laser illumination to destroy tumors while sparing adjacent tissue.

In scientific research, almost no field of physical or life science is free of lasers. Laser cooling uses photon momentum to slow atoms to within a billionth of a degree of absolute zero, enabling atomic physics and quantum simulation. Laser tweezers trap and manipulate individual molecules using radiation pressure, work that earned Arthur Ashkin the Nobel Prize in Physics in 2018. Fluorescence microscopy uses lasers to excite molecular labels in biological tissue, allowing biologists to visualize individual proteins inside living cells.

In telecommunications, semiconductor diode lasers modulated at billions of cycles per second carry the internet. Every video stream and financial transaction traveling through fiber infrastructure is encoded as pulses of laser light. The laser converted the global network from a copper-wire system of limited capacity into the photonic highway that carries virtually all of the world's digital traffic.

Diagram 4.11 - Industrial Laser Cutting Steel Sheet

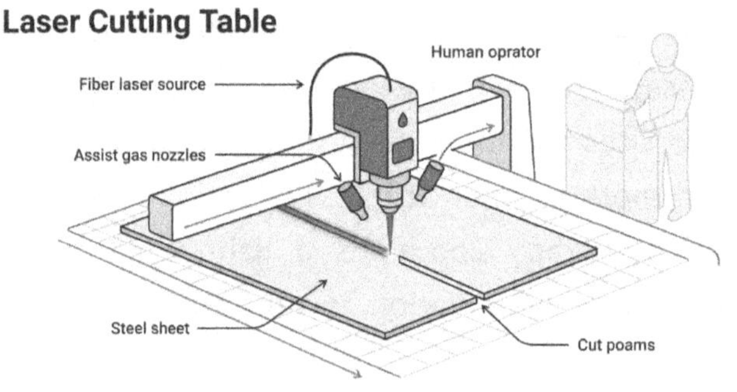

Kilowatt fiber lasers cut metal with micrometer precision, replacing mechanical tools across modern manufacturing.

The Optical Atomic Clock: Where Laser Precision Meets the Definition of Time

Of all laser applications, perhaps none is more philosophically striking than the optical atomic clock. An atomic clock uses the resonance frequency of an atom as its timekeeping oscillator. Since 1967, the cesium microwave transition has defined the international second. The best cesium fountain clocks toss cold atoms upward through a microwave resonator and measure the transition frequency as the atoms fall back down, achieving accuracy of about one part in ten to the fifteenth power. They would lose a second only after roughly thirty million years.

Optical atomic clocks go far beyond this by using optical transition frequencies, the transitions that produce visible or near-visible light, instead of microwave frequencies. Optical frequencies are roughly a hundred thousand times higher than microwave frequencies, and because more oscillations occur per second, the effective resolution is correspondingly finer. The challenge is reading out such a fast oscillation, which is where the frequency comb comes in. Modern optical atomic clocks based on strontium lattice systems achieve accuracies better than one part in ten to the eighteenth power. They are so precise that they can measure the gravitational redshift predicted by general relativity across a height difference of just a few centimeters between two laboratory benches.

Optical atomic clocks can map Earth's gravitational potential with better spatial resolution than any existing geodetic method. Any drift in the ratio of fundamental constants over cosmic time would appear as a measurable frequency shift when comparing two clocks based on different atomic species. They are also being developed as precision positioning systems for environments where satellite signals cannot reach.

Diagram 4.12 - Optical Atomic Clock with Cesium Fountain and Frequency Comb

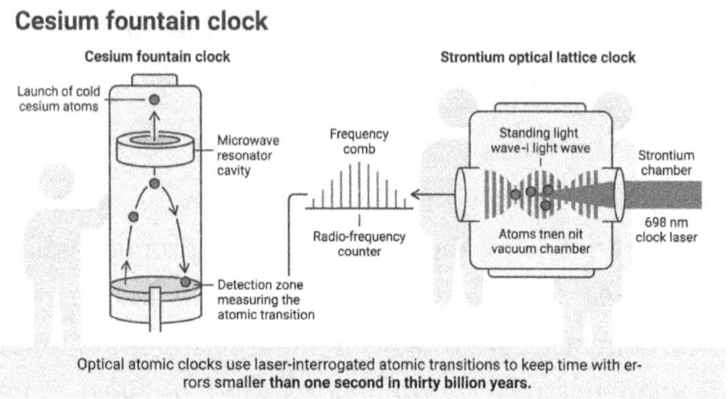

What This Means For You

In photonics, engineering, and any field using optical instrumentation, the laser is your primary tool, whether you think of it that way or not. Fiber optic transmitters are semiconductor diode lasers. Lidar rangefinders are Q-switched solid-state lasers. Pharmaceutical spectroscopic analyzers use precisely tuned diode or fiber lasers to probe absorption spectra.

For data infrastructure professionals, the terabits flowing through fiber networks depend on the coherence length and modulation bandwidth of semiconductor diode lasers. Wavelength-division multiplexed systems carrying dozens of simultaneous channels required narrow-linewidth lasers to avoid crosstalk, and future capacity growth depends on continued advances in coherence and energy efficiency.

For quantum technology practitioners, the laser is the central enabling tool: quantum key distribution uses attenuated laser sources to generate quantum states; trapped-ion quantum computers use precisely tuned beams to manipulate individual qubits; photonic quantum computers use laser-driven nonlinear processes to generate entangled photon pairs. In every case, coherence, monochromaticity, and phase control are not background knowledge but the conceptual foundation of the field.

Takeaway

Einstein's 1917 prediction of stimulated emission sat unused for nearly four decades. When engineers finally exploited it in 1960, the device was dismissed as a solution looking for a problem. Within a decade, it was cutting diamonds; within two, reading barcodes; within three, carrying the internet; and within four, correcting vision in twenty seconds without a blade. Few technologies better illustrate what happens when a quantum-mechanical principle is given the engineering tools to express itself.

The core principle is worth keeping in mind: laser light differs from ordinary light in three fundamental ways. It is monochromatic, consisting essentially of a single wavelength. It is coherent, with all photons marching in step over a long coherence length. And it is directional, propagating as a nearly parallel beam with almost no angular spread. All three properties arise from the same mechanism: stimulated emission in a population-inverted gain medium within an optical cavity that selects a single longitudinal mode and amplifies it relentlessly until a controlled fraction escapes as the output beam.

These properties make the laser the workhorse of modern photonics. In the next chapter, we follow laser light into the world's most important photonic infrastructure: the fiber optic cable, where a hair-thin strand of glass carries terabits per second across an ocean floor, pushed to limits that would have seemed fantastical to the scientists who first assembled a ruby rod between two mirrors in 1960.

5 Fiber Optics: The Photonic Highway

Opening Scenario

It is Tuesday evening, and you are on the couch. You tap Netflix and a two-hour film in crisp high definition begins within seconds. The Wi-Fi router across the room sends a radio pulse to your phone, carrying a fragment of the film. But the router is not the beginning of the story. A cable runs from it to the small box your internet provider installed near the front door, and from there, a fiber optic strand, a thread of glass thinner than a human hair, runs under the street to a local distribution node. From that node, more fiber reaches a regional data center full of humming server racks, linked in turn to metropolitan fiber rings, to internet exchange points, and ultimately to the backbone of the global internet.

That backbone, for most international traffic, is not a satellite in orbit. It is a cable on the ocean floor. Somewhere beneath the Atlantic, a bundle of glass strands no thicker than a garden hose carries the traffic of millions of simultaneous streams, video calls, and financial transactions. The film you just started watching almost certainly crossed that ocean as a sequence of light pulses traveling at roughly two-thirds the speed of light through the glass, covering the distance from Virginia to Ireland in about forty milliseconds. By the time the opening credits roll, those photons have already arrived, been decoded, and the pixels painted on your screen.

This chapter is the story of how that pipe works: a story about glass, the strange behavior of light at the boundary between two transparent materials, and the engineering decisions made over five decades that turned a laboratory curiosity into the infrastructure of civilization.

Why It Matters

Before fiber optics, long-distance telecommunications relied on copper wire, a medium that carries electrical signals and suffers serious signal degradation over modest distances. The physics imposed hard limits on how much information copper could carry. A transcontinental copper cable could handle thousands of telephone calls simultaneously if you were careful and clever. A single modern fiber strand can carry millions of simultaneous high-definition video streams. The difference is not incremental; it is a transformation of an entirely different order of magnitude.

That transformation affects everything you do with the internet: a phone call across a national border, a file retrieved from a cloud server, a video conference with someone on another continent. Fiber optics is also the substrate on which quantum communication networks are beginning to be built. Understanding the classical fiber first is the necessary foundation for the quantum layer that researchers are now constructing on top of it.

Total Internal Reflection: Light That Cannot Escape

The central puzzle of fiber optics is this: how does light travel for hundreds of kilometers through a glass strand without leaking out the sides? Glass is transparent. You can see through a window. So why does light not simply pass through the walls of the fiber?

The answer is total internal reflection. When light moves from one transparent material to another, it bends, an effect called refraction. The amount it bends depends on the refractive index, a number describing how much a material slows light relative to its speed in a vacuum. When light travels from a denser material, such as glass, into a less dense one, such as air, it bends away from the perpendicular. Increase the angle enough, and you reach the critical angle at which the refracted ray would have to travel exactly along the surface. Beyond that angle, there is no refracted ray at all: every bit of the light bounces back into the denser material. This is total internal reflection.

An optical fiber exploits this with precision. Its central glass core has a higher refractive index than the surrounding glass layer, called the cladding. Light launched into the core at the right angle strikes the core-cladding boundary at an angle steeper than the critical angle, and total internal reflection sends it back into the core. The light zigzags down the fiber, bouncing from wall to wall, with no energy leaking out. The numerical aperture of a fiber, derived from the refractive index difference between core and cladding, describes the cone of launch angles within which this confinement works. A higher numerical aperture allows light to enter more easily but can introduce signal distortion, as discussed shortly.

The purity of the glass is critical. Any imperfection, a microscopic bubble or impurity atom, can scatter light out of the core before it reaches the far end. Modern telecommunications fiber is made from silica glass refined by a process called modified chemical vapor deposition, which grows glass from chemical precursors in a controlled environment. The resulting glass blank, a preform, is then drawn into fiber at precisely controlled temperature and speed, preserving the core-cladding geometry over kilometer after kilometer of strand.

Diagram 5.1 - Total Internal Reflection in an Optical Fiber

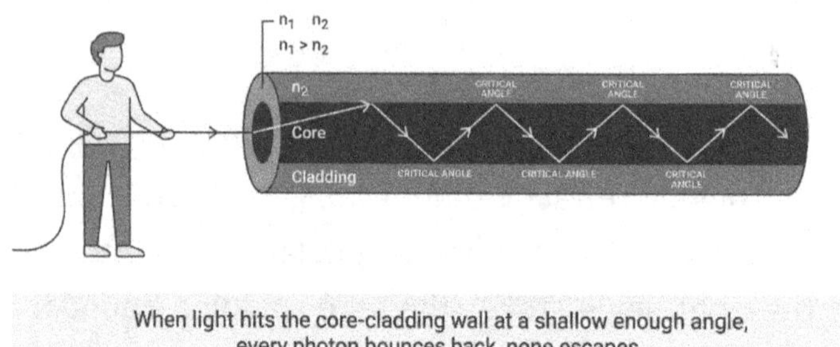

When light hits the core-cladding wall at a shallow enough angle, every photon bounces back, none escapes.

The Anatomy of a Fiber: Core, Cladding, and Jacket

Cut a telecommunications fiber and look at the end under a microscope, and you will see concentric rings. At the center is the core, the light-carrying region. Surrounding the core is the cladding. Outside the cladding is a protective polymer layer called the jacket.

The core is where light actually travels. In a standard single-mode telecommunications fiber, the core is approximately eight to nine micrometers in diameter; one micrometer is one millionth of a meter, making the core roughly ten times smaller than a human red blood cell. In multimode fibers, used for shorter distances, the core is larger, typically fifty to sixty-two and a half micrometers, making it easier to couple light from a source but introducing trade-offs in signal quality over long distances.

The cladding surrounds the core, with a refractive index slightly lower than the core's, typically by a fraction of a percent. That small but precisely controlled difference creates the boundary needed for total internal reflection. The refractive index profile, a graph showing how the refractive index varies from the core center outward, is one of the most important design parameters in fiber engineering. A step-index fiber has a uniform refractive index throughout the core that drops sharply at the cladding boundary. A graded-index fiber has a refractive index that is highest at the center and decreases smoothly outward, following a parabolic curve, which partially compensates for a distortion effect described in the next section.

Outside the cladding is the jacket, a polymer coating typically made of acrylate that protects the glass from moisture and abrasion. In submarine cables, additional layers of water-blocking compounds, copper conductors carrying electrical power to the amplifiers, layers of steel armor, and a final polyethylene sheath surround the fiber bundle. The result is a cable robust enough to withstand ocean-floor pressures, ship-anchor damage, and decades of continuous operation, while the glass threads inside remain protected.

Diagram 5.2 - Optical Fiber Cross-Section with Refractive Index Profile

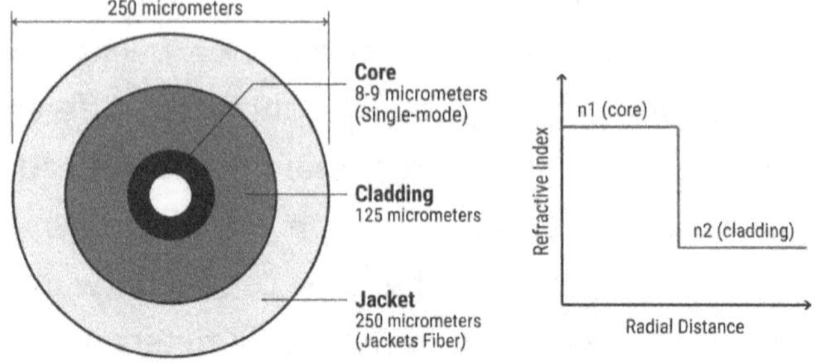

Three concentric layers, the narrowest thinner than a red blood cell, guide light for thousands of kilometers.

Single-Mode Versus Multimode: One Path or Many

In a fiber with a sufficiently large core, light can propagate along many different paths simultaneously. These different paths are called modes, and the phenomenon of multiple paths coexisting in the same fiber is called multimode propagation. Each mode travels at a slightly different effective speed because rays that bounce at steeper angles cover a greater total distance per unit of fiber length. A short pulse launched into a multimode fiber splits into many modes that arrive spread out in time at the far end. This temporal spreading is called modal dispersion, and it limits the data rate. If pulses spread too much, they overlap, and the receiver cannot distinguish individual bits.

Modal dispersion is acceptable for short distances. In a data center, where fiber links span tens to hundreds of meters, multimode fiber works well. Its larger core makes it easier to couple light from a source, reducing the cost of connectors and transceivers. For long-distance transmission, however, modal dispersion is crippling. The solution is single-mode fiber: a fiber with a core so narrow, approximately eight to nine micrometers, that only a single propagation mode can exist. When the core is this small relative to the wavelength of light, the geometry of the waveguide cannot support higher-order modes. Light travels along a single coherent path, eliminating modal dispersion.

A second distortion effect, chromatic dispersion, also called group velocity dispersion, affects single-mode fibers at very high speeds. Different wavelengths of light travel through glass at slightly different speeds. A real laser pulse contains a narrow but nonzero spread of wavelengths, and those wavelengths arrive at the far end spread in time even without modal dispersion. Engineers address chromatic dispersion through dispersion-shifted fiber, in which the refractive index profile is engineered to move the minimum-dispersion wavelength to the 1550-nanometer low-loss window, and through dispersion compensation modules inserted periodically along the link.

Diagram 5.3 - Single-Mode Versus Multimode Fiber Comparison

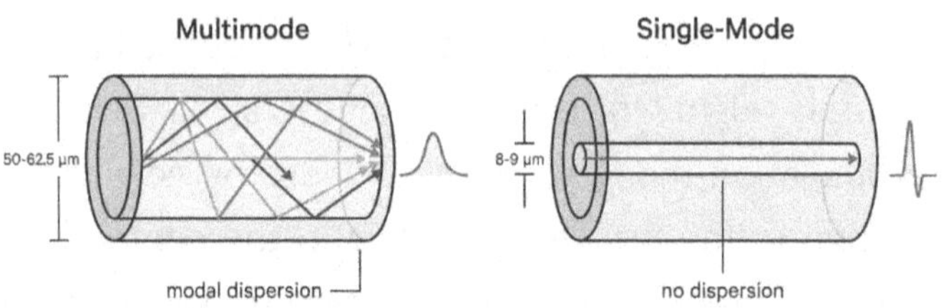

A narrower core forces light into one path, keeping data pulses sharp over thousands of kilometers.

Attenuation: Why Light Weakens, and How Engineers Measure It

Even the purest glass gradually absorbs and scatters light as it travels, reducing signal power over distance. This weakening is called attenuation and is measured in decibels per kilometer (dB/km). The decibel is a logarithmic unit: a loss of 3 dB means signal power is halved, 10 dB means it drops to one-tenth, and 20 dB means it drops to one-hundredth. Over long distances, these losses accumulate, so that a transatlantic link without amplification would reduce the signal to a power level so vanishingly small that any conceivable receiver could not detect it.

Two physical processes dominate attenuation in silica fiber. Rayleigh scattering occurs because glass, even when cooled from the melt, retains random density fluctuations at the atomic scale that scatter light in all directions. Scattering increases strongly at shorter wavelengths, which is the same effect that makes the sky blue.

Absorption by trace impurities or by the silica lattice itself at certain wavelengths adds a second contribution. Early fiber exhibited a pronounced absorption peak near 1380 nanometers, caused by hydroxyl ions from water contamination; modern manufacturing has reduced this to negligible levels.

The combined effect produces a characteristic attenuation curve: high at short wavelengths, falling steeply as wavelength increases, reaching a minimum in the near-infrared, then rising again. There are two key low-loss windows. The 1310-nanometer window has an attenuation of roughly 0.35 dB/km and offers very low chromatic dispersion in standard single-mode fiber. The 1550-nanometer window has the lowest attenuation of all, roughly 0.18-0.20 dB/km, and is the primary window for long-haul and submarine telecommunications. The entire ecosystem of long-haul fiber technology, laser wavelengths, amplifiers, and wavelength channel allocations is built around this window.

Diagram 5.4 - Fiber Attenuation Curve with 1310nm and 1550nm Windows

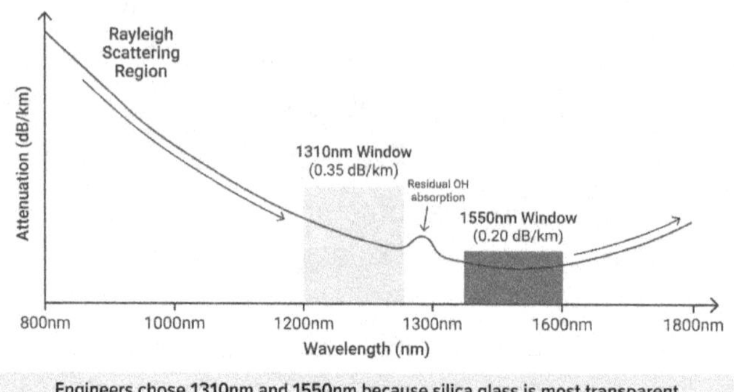

Engineers chose 1310nm and 1550nm because silica glass is most transparent at these near-infrared wavelengths.

To appreciate why amplification is essential, consider a transatlantic link spanning 6,000 kilometers. At 0.20 dB/km, the accumulated attenuation is 1,200 dB, reducing signal power by a factor of 10 to the 120th power. The number of atoms in the observable universe is around 10 to the power of 80. No detector could find a signal that faint. Optical amplifiers placed at intervals of roughly 50 to 80 kilometers solve this by periodically restoring the signal to its original level.

Erbium-Doped Fiber Amplifiers: Boosting Light Without Converting It

For the first two decades of the fiber era, long-haul links used regenerators: devices that converted the optical signal to an electrical signal, amplified it electronically, then converted it back to light. Regenerators were expensive, speed-specific, and wavelength-specific. Upgrading the capacity of a submarine cable meant replacing regenerators at every amplification point, including those sitting on the ocean floor several kilometers below the surface.

The erbium-doped fiber amplifier, or EDFA, demonstrated in the late 1980s and deployed in submarine systems in the early 1990s, changed everything. The EDFA amplifies optical signals directly in the fiber without any conversion to electrical form. A short section of optical fiber, typically a few meters to tens of meters, is fabricated with erbium ions incorporated into the glass of the core. Erbium is a rare-earth element whose ions, when they absorb photons at pump wavelengths of 980 or 1480 nanometers, reach an excited energy state and remain there for several milliseconds before emitting a photon and returning to the ground state. This long lifetime allows a population of excited ions to build up in the fiber.

When a signal photon in the 1550-nanometer window passes through the doped section, it can stimulate an excited erbium ion to emit an additional photon at the same wavelength, the same direction, and the same phase as the signal. This is stimulated emission: one photon in, two photons out. The energy comes from the pump laser that keeps the erbium ions excited. The signal emerges from the doped section stronger than it entered, with amplification supplied entirely in the optical domain.

The most important property of the EDFA is wavelength independence. The gain band spans roughly 1530 to 1610 nanometers, covering the entire C-band and L-band. Any signal wavelength within the band is amplified simultaneously. Adding a new wavelength channel requires only a new laser at the endpoint; the EDFAs along the route amplify it automatically alongside all existing channels. This property directly enabled the wavelength-division multiplexing revolution described in the next section.

Diagram 5.5 - EDFA Erbium-Doped Fiber Amplifier Schematic

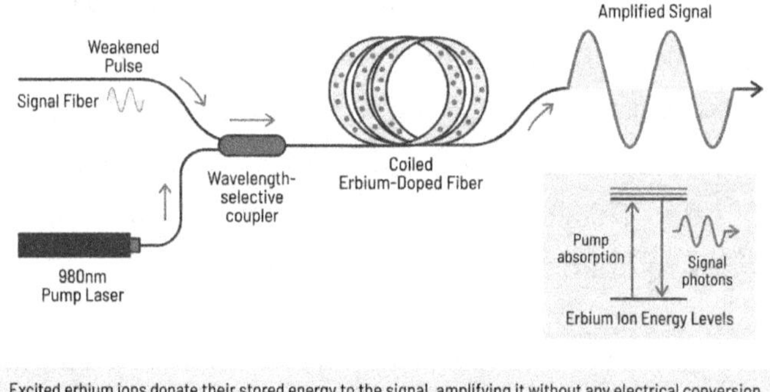

Excited erbium ions donate their stored energy to the signal, amplifying it without any electrical conversion.

Wavelength-Division Multiplexing: Many Colors, One Fiber

Light has a property that makes it uniquely suited to carrying enormous amounts of information: two beams at different wavelengths can coexist in the same fiber without interfering, each carrying its own independent data stream. Wavelength-division multiplexing, or WDM, exploits this by combining multiple laser sources, each emitting at a slightly different wavelength, onto the same fiber using a device called a multiplexer (mux). At the receiving end, a demultiplexer, or demux, separates the combined beam back into individual wavelength channels and routes each to its own receiver. The link behaves like many parallel fiber connections, even though it uses a single physical strand.

Dense wavelength-division multiplexing, or DWDM, packs wavelength channels much more closely together than early coarse WDM systems. The International Telecommunication Union has standardized DWDM channel spacings of 100, 50, and 25 gigahertz, corresponding to wavelength separations of roughly 0.8, 0.4, and 0.2 nanometers at the 1550-nanometer center wavelength. Standard DWDM systems deploy 40, 80, or 96 channels within the EDFA's C-band amplification window. At 80 channels each carrying 100 gigabits per second, a single fiber reaches 8 terabits per second in aggregate. Modern coherent optical systems using advanced modulation formats push individual channels to 400 gigabits per second or one terabit per second, placing total fiber capacity close to 100 terabits per second on a single strand.

Diagram 5.6 - WDM/DWDM Multiplexer Combining Many Wavelengths

Wavelength Division Multiplexing (WDM)

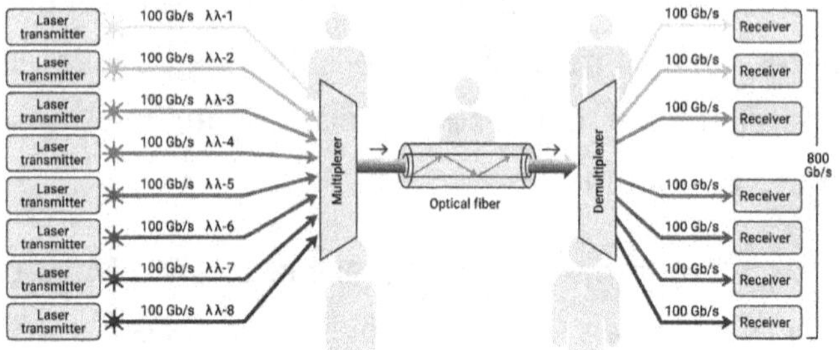

Each laser color carries its own data stream; the multiplexer weaves all colors into one fiber while the demultiplexer separates them at the far end.

The arrayed waveguide grating, or AWG, is a common multiplexing device that uses waveguide paths of precisely graded lengths to sort wavelengths into different output ports without electronics or a power supply. In modern metropolitan networks, reconfigurable optical add-drop multiplexers, or ROADMs, allow specific wavelength channels to be extracted for local delivery and replaced with fresh traffic under software control, without disturbing passing channels. ROADMs let operators reroute traffic and reconfigure capacity remotely.

The Global Fiber Network: Cables Under the Ocean and Fiber to Your Door

Approximately 99 percent of all international internet and telecommunications traffic travels through submarine fiber-optic cables on the ocean floor, not through satellites. Satellites reach remote locations, ships at sea, and aircraft. Still, for backbone traffic between continents, submarine cables carry the load: they offer far higher capacity, lower latency, and lower cost per transmitted bit.

More than 400 active submarine cable systems, spanning over 1.3 million kilometers, connect every inhabited continent. Ocean floor surveys select routes avoiding canyons and fishing grounds. The cable is manufactured in factories, loaded onto purpose-built cable ships, and paid out as the ship follows the route. EDFA repeater housings are spliced in every 50 to 80 kilometers, each in a pressure vessel rated for ocean depths up to 8,000 meters, powered by constant electrical current through copper conductors from landing stations on shore.

Diagram 5.7 - Transatlantic Submarine Cable on Ocean Map

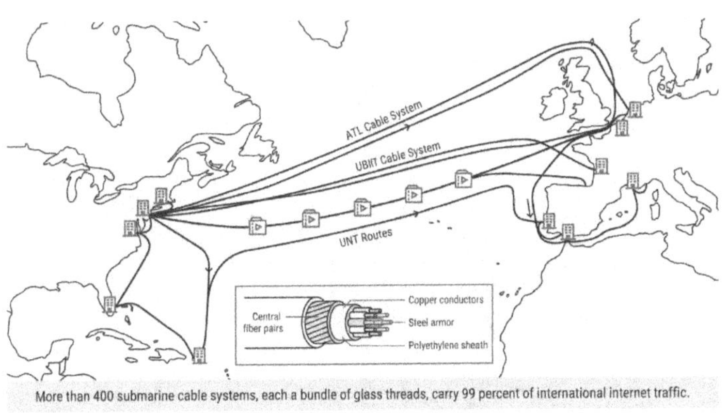

More than 400 submarine cable systems, each a bundle of glass threads, carry 99 percent of international internet traffic.

On land, a hierarchy of fiber networks brings capacity from the submarine cable landing stations to individual users. Long-haul backbone links connect major cities and data center clusters using high-capacity DWDM. Metropolitan area networks link data centers, exchange points, and large business customers within a city. Access networks connect individual buildings and homes. The endpoint for millions of residential customers is fiber to the home, or FTTH, sometimes called fiber to the premises. In an FTTH deployment, a fiber strand runs from the service provider's exchange to the customer's building. Inside, an optical network terminal, a small box that converts the optical signal to electrical for the Wi-Fi router and devices, provides the final connection. The practical experience is simply a fast internet connection; the underlying reality is photons traveling from a distant data center to within meters of your device before the signal is converted into anything other than light.

Diagram 5.8 - Data Center Server Racks and Internal Fiber Network

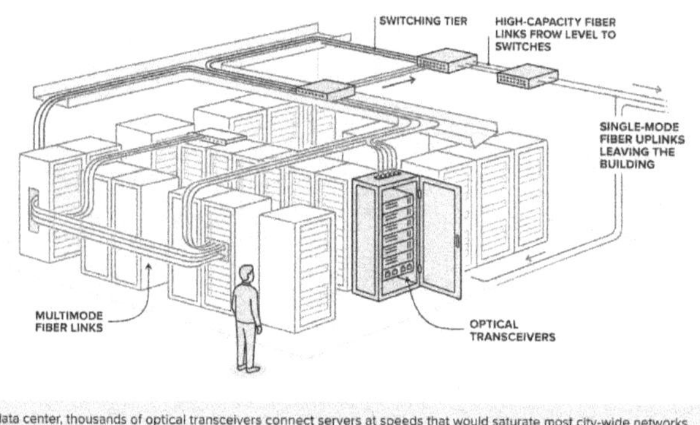

Inside a data center, thousands of optical transceivers connect servers at speeds that would saturate most city-wide networks.

Measuring and Maintaining the Network: OTDR and Fiber Splicing

A fault in a fiber optic network is invisible to the naked eye. The fiber may look intact yet carry no light at all because the damage is inside the glass: a microbend, a broken strand, a poorly made splice, or a contaminated connector. The primary instrument for locating faults is the optical time-domain reflectometer, or OTDR.

The OTDR sends a short, intense pulse of light into one end of the fiber and measures the light that returns. As the pulse travels, a small fraction scatters backward toward the source through Rayleigh scattering. This backscattered light returns at a rate directly related to distance: light scattered 10 kilometers away returns after the time it takes to travel 20 kilometers through the glass, roughly 100 microseconds. By measuring backscattered power over time, the OTDR constructs a profile of the fiber along its entire length. A connector or splice appears as a step down in the profile; a fiber break appears as a sudden termination with a bright reflection spike. An OTDR can locate a fault to within a meter over a link spanning many kilometers.

Diagram 5.9 - OTDR Fault Location Display

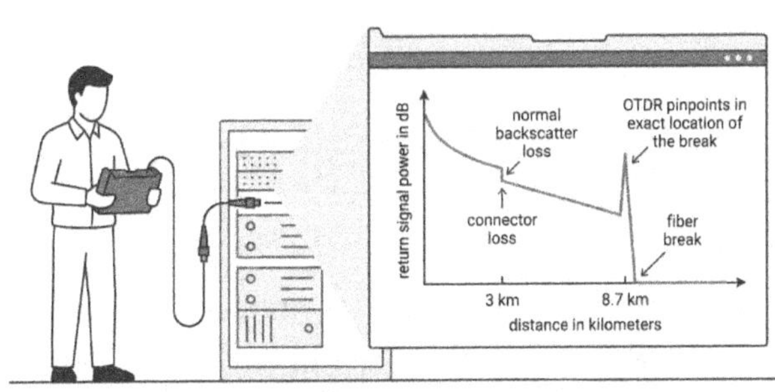

By timing the return of backscattered light, an OTDR pinpoints a fiber fault within a meter over a link spanning many kilometers.

When a fault is found, it is repaired by splicing: joining two fiber ends to create a continuous, low-loss optical path. Mechanical splices use a precision sleeve and a refractive-index-matching gel, producing a loss of 0.1 to 0.3 dB. Fusion splices use a fusion splicer, which fires a brief electric arc to melt and fuse two fiber ends after positioning them to sub-micrometer precision with motorized alignment stages. A well-made fusion splice has a loss of 0.01 to 0.02 dB, essentially undetectable, and the joint is nearly as strong as the original fiber. Modern splicers complete a fusion splice in under one minute.

Diagram 5.10 - Fiber Fusion Splice Machine

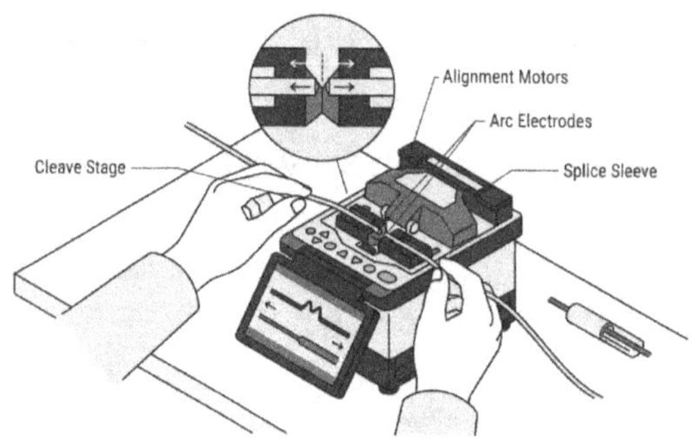

A fusion splicer melts two fiber ends together in seconds, creating a joint nearly as strong and transparent as the original fiber.

Fiber to the Home: The Last Mile and What It Means

The phrase last mile refers to the connection between the service provider's nearest distribution point and the individual customer's premises. For decades, this last mile was the weakest link in the global network: transoceanic fiber cables carried terabits per second, but the final connection to the home was copper telephone wire from the nineteenth century, capable of only a few megabits per second. Fiber to the home, or FTTH, replaces this copper link with fiber, extending the photonic highway all the way to individual homes or apartments.

The dominant FTTH architecture is the passive optical network, or PON. In a PON, a single optical fiber runs from the telephone exchange to a passive optical splitter in the neighborhood, from which a tree structure fans out to individual homes. The keyword is passive: the splitter contains no electronics, no power supply, and no active components. Light from the exchange passes through the splitter and is distributed to all connected homes simultaneously. Electronics are located only at the exchange and at each customer's premises, reducing costs and eliminating the maintenance burden of active equipment in the street.

Diagram 5.11 - FTTH Fiber to the Home Installation Architecture

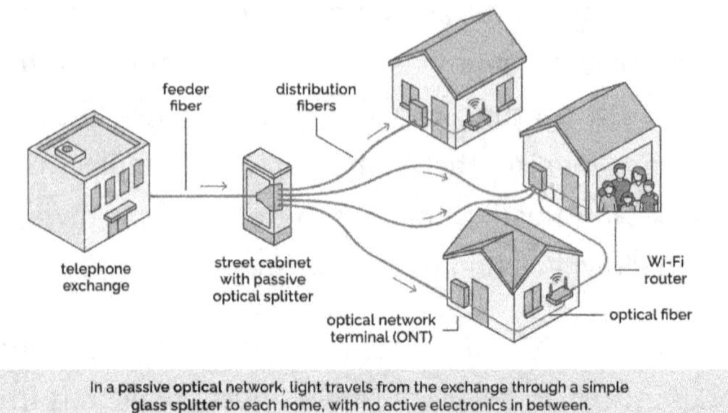

In a **passive optical** network, light travels from the exchange through a simple **glass splitter** to each home, with no active electronics in between.

FTTH connections routinely deliver symmetrical speeds of 500 megabits per second to several gigabits per second. A copper wire degrades with distance, humidity, and age; fiber does not. A fiber installed today can carry far more data tomorrow by upgrading only the transceivers at the endpoints. The most cost-intensive part of an FTTH rollout is not the fiber but the civil engineering: digging trenches and boring under driveways. Micro-trenching, cutting a narrow slot a few centimeters wide in the road surface, has significantly reduced deployment costs in urban environments.

What This Means For You

If you work in network engineering, systems architecture, or cloud infrastructure, fiber optics is the physical substrate on which almost everything your work depends. The difference between single-mode and multimode fiber determines purchasing decisions for data center interconnects. Attenuation budgets and dispersion limits govern whether a proposed fiber link will perform reliably at the data rates your applications require. DWDM capacity planning is directly relevant to any role involving provisioning or scaling network bandwidth. These concepts are the vocabulary of the physical layer, and the physical layer is the one that cannot be abstracted away when something breaks.

For those working in quantum information and quantum communications, the connection is direct. Quantum key distribution networks, which encode cryptographic key information in the quantum states of individual photons, operate over fiber-optic infrastructure. The properties that matter for quantum communication, attenuation, dispersion, and polarization-mode dispersion, are properties of the same fiber discussed in this chapter. Researchers designing city-scale quantum networks work with the same single-mode fiber, the same connector losses, and many of the same measurement tools, including the OTDR, that classical engineers have used for decades. The classical fiber layer is a prerequisite for anyone engaging seriously with quantum networking.

For policymakers and security professionals, submarine cables are a strategic infrastructure. A relatively small number of cables carry virtually all internet traffic between continents, and deliberate damage to even a few strategically selected cables could severely disrupt communications across entire regions. Several incidents in recent years have severed submarine cables, causing measurable disruptions to internet services in the affected areas. Professionals who understand this infrastructure are better equipped to assess risk, design resilient topologies, and evaluate the credibility of threat reports.

Takeaway

Fiber optics transformed global communications from a copper-limited system into the photonic highway that carries modern civilization's information traffic. The foundation is total internal reflection: light traveling inside a glass core with a slightly higher refractive index than the surrounding cladding cannot escape through the walls, bouncing forward kilometer after kilometer, confined by geometry and physics.

Single-mode fiber, with its eight-micrometer core, eliminates modal dispersion for intercontinental links. Multimode fiber, with its larger core, is well-suited to short data-center runs at lower cost. Attenuation reaches its minimum at 1550 nanometers, and the entire ecosystem of long-haul technology, from laser wavelengths to amplifier designs to DWDM channel allocations, is built around that window. EDFAs amplify signals across the full band without electrical conversion, enabling DWDM to pack dozens to hundreds of independent wavelength channels onto a single strand, collectively approaching 100 terabits per second.

The result is a global network of more than 400 submarine cable systems crossing every ocean, supplemented by metropolitan fiber rings and FTTH access networks that extend the photonic highway to individual homes. The OTDR locates faults by timing backscattered light; the fusion splicer repairs them with sub-micrometer precision; the passive optical splitter fans out capacity to many homes without active electronics in the street. These tools maintain an infrastructure on which the modern digital economy depends.

The next chapter moves deeper into the quantum nature of light. Everything so far has treated photons as classical light pulses traveling through a waveguide. But a single photon can exist in a superposition of polarization states, carry quantum information that cannot be copied without detection, and become entangled with another photon in ways that have no classical analog. That quantum layer, built on top of the classical fiber infrastructure explored here, is where the frontier of communications and computing is being pushed today.

Diagram 5.12 - Chapter Summary: From Living Room to Ocean Floor

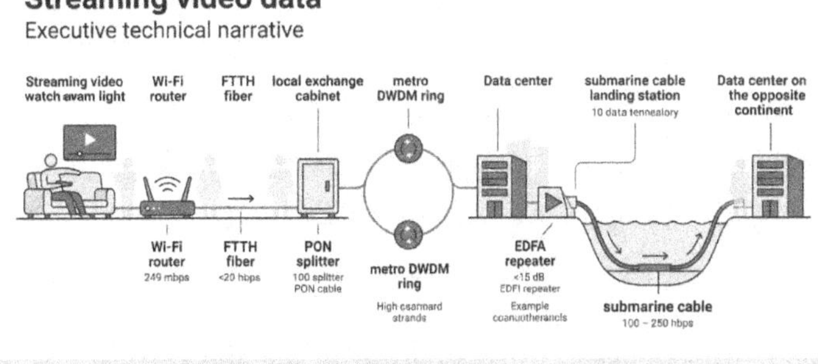

From couch to ocean floor, your stream travels as light pulses through a hierarchy of glass strands built over five decades of photonic engineering.

6 Quantum States of Light: Beyond Classical Waves

Opening Scenario

You settle into a corner seat at a busy coffee shop, slide noise-canceling headphones over your ears, and wait. The espresso machine hisses. A delivery truck rattles past the window. Then the headphones engage, and the roar drops away. But not completely. A faint, low hiss remains beneath everything, a residual noise that no amount of engineering can fully eliminate because the electronic circuits inside generate their own thermal agitation. You have hit the technology's noise floor, and there is no classical path around it.

Physicists working with light face an analogous problem at a far more fundamental level. Any measurement made with ordinary laser light, no matter how carefully the laser is designed, is ultimately limited by a background of quantum noise called shot noise. Shot noise is not the result of imperfect engineering. It is woven into the fabric of quantum mechanics itself, and for most of the history of precision optics, it was assumed to be an immovable floor. What quantum optics has revealed is that there is a way around it, not by ignoring the rules of quantum mechanics, but by using them more cleverly.

At LIGO, the Laser Interferometer Gravitational-Wave Observatory, scientists measure displacements smaller than one-thousandth the diameter of a proton across baselines four kilometers long. The signals they hunt, faint ripples in spacetime from colliding black holes a billion light-years away, would be completely washed out by quantum shot noise if left unaddressed. Since 2019, LIGO has used a technology called squeezed light, a quantum state in which the noise has been deliberately reshaped at the quantum level, to push beyond that floor. The result is the most sensitive measuring instrument ever built, and a vivid demonstration that quantum states of light are practical engineering resources, not academic curiosities. This chapter explains how all of that works.

Why It Matters

Earlier chapters described light as both a wave and a stream of photons, and explored how lasers and optical fibers exploit these properties to transmit information with extraordinary efficiency. All of that physics sits comfortably within the classical regime, where the quantum nature of light appears only as a background effect. Chapter 6 crosses a threshold. Beyond this threshold, the quantum mechanical structure of light is not a nuisance to minimize but a resource to engineer. Understanding quantum states of light is the conceptual gateway to quantum sensing, quantum communication, and photonic quantum computing.

The technologies described in this chapter already exist and already matter. Squeezed light is in active use at gravitational-wave detectors. Single-photon sources are being developed for quantum cryptography and quantum computing worldwide. Photon-number-resolving detectors are moving from research laboratories into commercial products. Engineers and physicists working in these fields, and the managers and strategists who work alongside them, need a clear picture of what a quantum state of light is, why different quantum states behave so differently, and why the distinctions matter for real instruments.

What a Quantum State of Light Actually Means

In classical physics, describing a beam of light is straightforward. You specify the amplitude of the electric field, its frequency, its polarization, and its phase, and you have a complete description. There is nothing uncertain about those quantities in the classical picture; they are fixed numbers you could, in principle, measure simultaneously with perfect precision, given good enough instruments.

Quantum optics tells a different story. A quantum state of light specifies the probability distribution for every possible measurement outcome: the probability of detecting zero photons, one photon, two photons; the probability of measuring a particular electric field amplitude; the probability of finding the field at a particular phase. These probabilities are not a reflection of ignorance that better measurements would remove. They are a fundamental feature of reality at the quantum level.

The electric field of a light beam can be decomposed into two components called quadratures, which you can picture as two independent axes in a mathematical space describing the field. For a classical wave, both quadratures have definite, knowable values. For a quantum state, each quadrature has its own spread of probable values, and the spreads are linked by a fundamental constraint: the Heisenberg uncertainty principle. You cannot simultaneously reduce the uncertainty in both quadratures below a certain minimum product. You can, however, push the uncertainty in one quadrature well below that minimum, provided you accept a compensating increase in the uncertainty of the other. That trade-off is the engine behind squeezed light.

Phase space is a useful visual tool for comparing quantum states. Imagine a flat diagram with the two quadratures plotted along horizontal and vertical axes. A classical state with a definite amplitude and phase is a single point. A quantum state is a blob of probability spread over some region. The shape and orientation of this blob distinguish different quantum states: a coherent state (laser light) is a symmetric circle; a squeezed state is an ellipse; a Fock state (a state of exactly defined photon number) produces a ring. These shapes encode everything about how the state behaves in measurements.

The practical significance of the phase-space picture is that two light beams with identical average intensities can behave completely differently in a precision measurement if they are in different quantum states. Intensity alone does not characterize a light state at the quantum level. The full quantum state, the complete probability distribution over phase space, determines what measurements are possible and how precisely they can be made. Classical optics did not need this level of description; quantum engineering cannot proceed without it.

Diagram 6.1 - Phase-Space Portraits of Three Quantum States

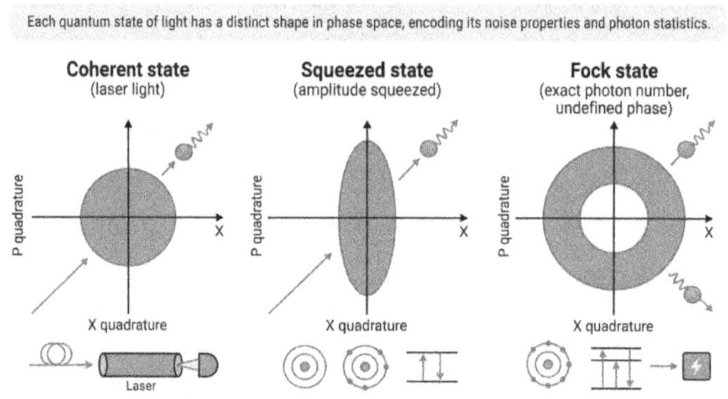

The Heisenberg Uncertainty Principle Applied to Photons

Most people first encounter the Heisenberg uncertainty principle in the context of particles: you cannot simultaneously know the exact position and the exact momentum of an electron. The uncertainty principle is far more general than this single example. It applies to any pair of quantities that are complementary in the mathematical sense, meaning that measuring one necessarily disturbs the other.

For light, the relevant complementary pairs are the two quadratures of the electromagnetic field. Call them the X quadrature and the P quadrature. They are linked so that the product of their uncertainties cannot fall below a fixed minimum. This minimum defines the shot-noise limit, also called the standard quantum limit. Shot noise refers to the random fluctuations caused by the discrete arrival of individual photons: even if the average arrival rate is perfectly steady, photons arrive at random intervals, creating irreducible fluctuations in any measurement. These fluctuations scale with the square root of the average photon count, and they set the noise floor for any measurement using classical laser light.

The shot-noise limit is not a technological limit that better engineering could overcome. It is set by nature. No matter how pure the laser or how quiet the electronics, you cannot reduce measurement noise below the shot-noise limit using classical light. What quantum mechanics does permit, however, is redistribution. The uncertainty principle constrains the product of the two quadrature uncertainties, but it says nothing about how that product is divided between them. You are free to push uncertainty from one quadrature into the other. If the quadrature you are measuring is the one you have squeezed down, your measurement noise is below the standard quantum limit.

You have not violated the uncertainty principle; you have exploited it.

A simple analogy helps here. Imagine carrying a fixed volume of water in a flexible balloon. The volume cannot shrink, just as the uncertainty product cannot fall below its minimum. Normally, the balloon is roughly spherical, equal in all directions. But you can squeeze it with your hands, making it elongated: narrow in one direction, wide in another. The volume stays the same; the shape changes. Squeezed light does exactly this to the quantum noise of a light beam, reshaping the uncertainty blob in phase space so that it is narrow in the measured direction and wide in the unmeasured direction.

This geometric picture clarifies a subtlety about the standard quantum limit: it is not the absolute minimum uncertainty allowed by quantum mechanics, but the minimum for states in which the uncertainty is symmetrically distributed between the two quadratures. Coherent states sit at exactly this symmetric minimum. Squeezed states break the symmetry deliberately, accepting more uncertainty in one direction to gain less in another. The Heisenberg principle is never violated; the noise budget is allocated more strategically. The terms shot noise and quantum noise are sometimes used interchangeably. Still, shot noise refers specifically to Poisson fluctuations from discrete photon arrivals, while quantum noise is the broader category that also includes vacuum fluctuations and radiation-pressure effects.

Diagram 6.2 - Heisenberg Uncertainty and the Shot-Noise Limit

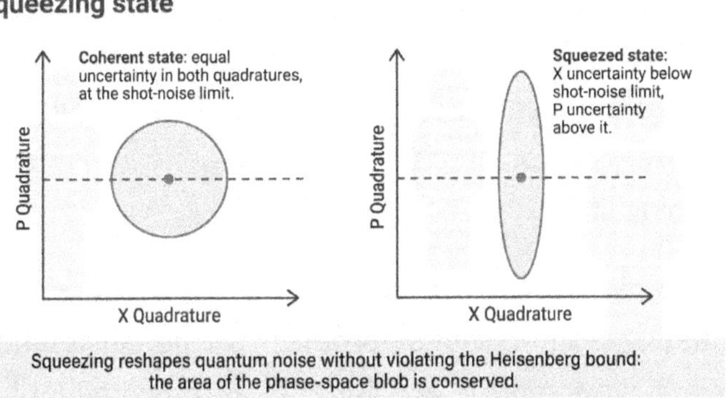

Coherent States: The Quantum Description of Laser Light

The quantum state that most closely resembles a classical electromagnetic wave is the coherent state, formalized by physicist Roy Glauber in the 1960s. In a coherent state, the photon number is not fixed. It fluctuates around an average value following the Poisson distribution, the same statistical pattern that describes random events occurring at a constant average rate, like buses arriving at a stop. The spread of those fluctuations equals the square root of the average photon number, and this spread is precisely the shot noise associated with laser light.

In phase space, the coherent state is a symmetric circular blob centered on a point that encodes the amplitude and phase of the corresponding classical wave. Both quadratures carry exactly half of the total minimum uncertainty, and neither is squeezed relative to the other. This is why laser light is described as 'as classical as quantum light can be': it is the quantum state with the most symmetrically distributed minimum uncertainty, looking as much like a smooth wave as quantum mechanics permits. But even a perfect laser is subject to shot noise, and for precision measurements, this noise is the binding limit.

Understanding the coherent state also reveals why attenuating a laser cannot produce a true single-photon source. If you reduce a coherent laser beam until the average photon number is one, the Poisson distribution means you still sometimes get zero photons, sometimes two, sometimes three. The photon number fluctuates. What quantum information applications need is a state with exactly one photon and no probability of getting two, which requires an entirely different physical platform. That is what a Fock state provides.

Diagram 6.3 - Coherent State Phase-Space Diagram

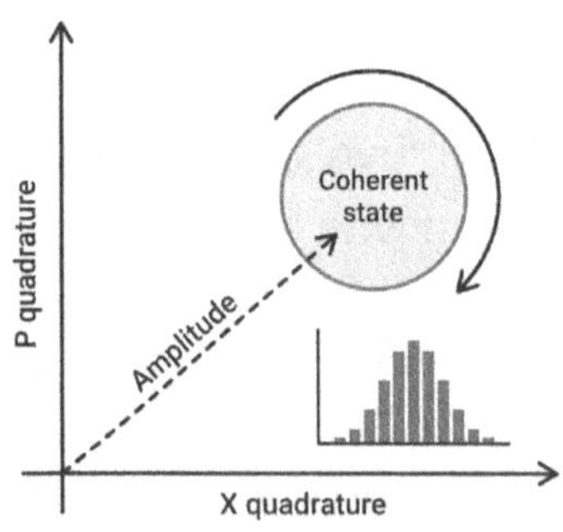

Laser light in a coherent state: symmetric noise, Poisson photon statistics, the most classical quantum light possible.

In quantum optics, the modes of the electromagnetic field behave mathematically like quantum harmonic oscillators. The coherent state is the quantum state of a harmonic oscillator that most closely tracks a classical oscillation, which explains why it arises naturally from a laser cavity, preserves its shape as it propagates, and is stable under many optical operations. All departures from the coherent state into the truly quantum regime require deliberate engineering of the light field.

Fock States: Light With Exactly Defined Photon Numbers

If the coherent state is quantum light behaving as classically as possible, the Fock state is quantum light at its most non-classical. A Fock state, also called a number state and named after Soviet physicist Vladimir Fock, is a quantum state in which the photon number is precisely defined. Not probabilistic, not fluctuating: exactly fixed. A state with exactly zero photons is the vacuum state, written |0>. A state with exactly one photon is written |1>. A state with exactly two photons is |2>. The photon-number ladder climbs from |0> upward through every positive integer.

The vacuum state, |0>, deserves special attention. It is a state with no photons, yet it is not empty in the classical sense. The Heisenberg uncertainty principle means that even with zero photons, the two quadratures of the field cannot both be simultaneously zero. These residual fluctuations, called vacuum fluctuations, are physically real: they contribute to spontaneous emission from excited atoms, generate the Casimir force between closely spaced metal plates, and seed the parametric processes that produce squeezed light. The quantum vacuum is nothing; it is a substrate of irreducible zero-point energy.

Diagram 6.4 - The Photon-Number Ladder: Fock States

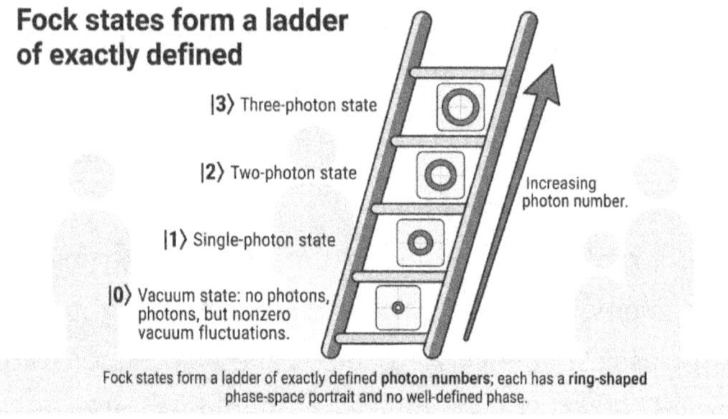

Fock states form a ladder of exactly defined **photon numbers**; each has a **ring-shaped** phase-space portrait and no well-defined phase.

In phase space, a Fock state with n photons produces a ring-shaped probability distribution: the photons are present with a definite count, but the phase of the field is completely undefined. This complementarity between photon number and phase is a direct consequence of the uncertainty principle. A Fock state represents the extreme of maximum photon-number precision and maximum phase uncertainty.

The single-photon Fock state, |1>, is the fundamental resource for quantum cryptography, quantum computing, and quantum communication. A single-photon source is a device that emits exactly one photon per trigger event.

Quantum dots (nanoscale semiconductor crystals that confine a single electron) emit one photon as the electron returns to its ground state after a laser pulse. Nitrogen-vacancy centers in diamond act as artificial atoms at the atomic scale and emit single photons with high purity. Nonlinear optical processes can also serve as heralded sources: detecting one photon from a correlated pair announces the presence of its partner.

Diagram 6.5 - Single-Photon Source Schematic

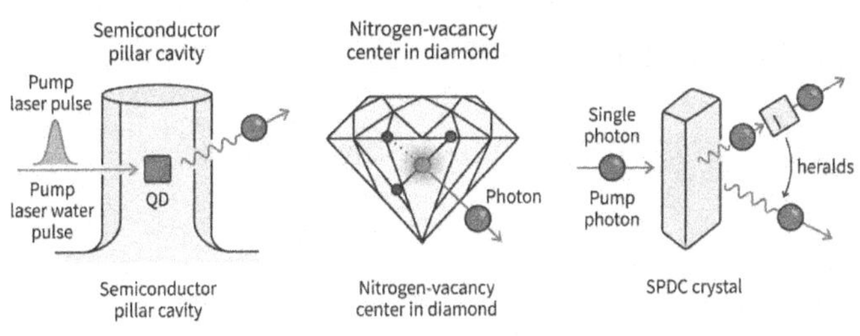

True single-photon sources use quantum emitters; an attenuated laser cannot produce a genuine Fock state |1>.

Detecting Fock states precisely requires a photon-number-resolving detector, often abbreviated PNR. Standard single-photon avalanche diodes can detect the presence or absence of light but cannot distinguish between one photon and two: they produce the same click for any nonzero photon number. Superconducting nanowire PNR detectors, cooled to temperatures near absolute zero, can reliably distinguish photon numbers up to twenty or thirty, enabling precise characterization of quantum light sources and states in quantum computing laboratories.

The engineering challenge of a single-photon source is not simply to produce Fock states but to produce them on demand, at high rates, with high purity, and with photons that are indistinguishable from each other. Indistinguishability, meaning that two photons are identical in frequency, polarization, and temporal shape, is essential for quantum computing applications that require multiple photons to interfere at a beam splitter. Current semiconductor quantum dot sources can approach 99% purity and 98% indistinguishability under optimized low-temperature conditions, but achieving both simultaneously at practical repetition rates and elevated temperatures remains an active research challenge.

Diagram 6.6 - Photon-Number-Resolving Detector

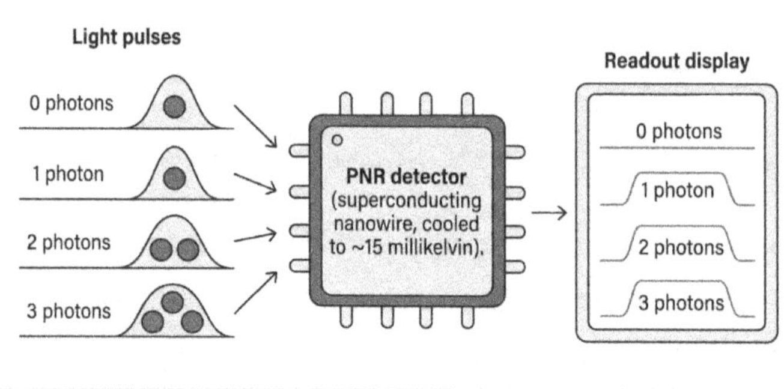

Photon-number-resolving detectors distinguish exact photon counts, unlike standard detectors that only signal the presence or absence of light.

Squeezed Light: Engineering the Shape of Quantum Noise

Squeezed light is the quantum state in which the uncertainty in one quadrature has been deliberately reduced below the shot-noise limit. In contrast, the uncertainty in the complementary quadrature is correspondingly increased. The total uncertainty product respects the Heisenberg bound, but its distribution has been reshaped. If the squeezed quadrature is the one being measured in a precision experiment, the measurement noise is reduced below anything a classical laser could achieve.

The physical process that produces squeezed light is called optical parametric amplification, most commonly implemented via parametric down-conversion. In a specially engineered nonlinear crystal, a material whose optical properties change with light intensity, a single high-energy pump photon is converted into two lower-energy photons called the signal and idler. These two photons are quantum-correlated; their joint quantum state encodes correlations that, when properly harvested, yield squeezed noise in the output light. The crystal is typically placed inside an optical cavity, a pair of mirrors that cause the light to pass through the crystal repeatedly, building up the squeezing with each pass. This arrangement is called an optical parametric oscillator, or OPO.

Diagram 6.7 - Parametric Down-Conversion Crystal Generating Squeezed Light

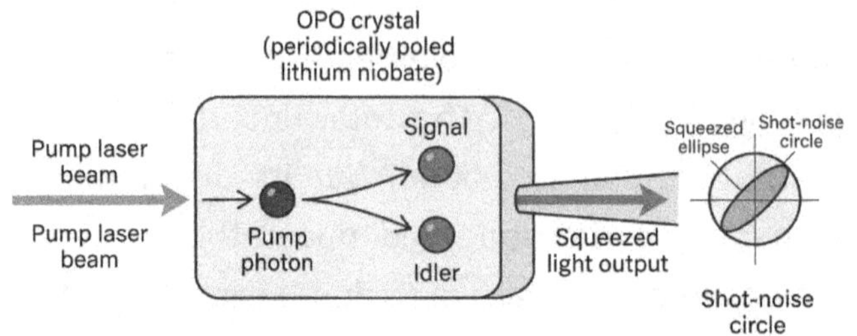

Parametric down-conversion converts pump photons into quantum-correlated pairs, reshaping the noise into a squeezed ellipse.

The degree of squeezing is measured in decibels relative to the shot-noise floor. Ten decibels of squeezing means the noise power in the squeezed quadrature is ten times smaller than the shot-noise floor, corresponding to roughly a factor-of-three improvement in signal-to-noise ratio. Current state-of-the-art laboratory squeezers achieve 15 dB or more under optimized conditions; 10-12 dB is routine in well-engineered systems. The main practical challenge is optical loss, which degrades squeezing by mixing the squeezed state with ordinary vacuum. Even a few percent loss can significantly reduce usable squeezing, making low-loss optical design a central priority.

Squeezed light comes in several varieties. Amplitude-squeezed light reduces noise in the amplitude quadrature, improving photon-counting measurements at the cost of increased phase noise. Phase-squeezed light does the reverse. Two-mode squeezed light involves correlations between two separate beams, the signal and idler outputs of a parametric process, making it especially useful for quantum communication protocols. All varieties share the same origin: redistribution of quantum noise by a nonlinear optical process, always constrained by the Heisenberg bound but shaped freely within it.

A key practical challenge for any system using squeezed light is maintaining the correct phase relationship between the squeezer and the measurement apparatus. Squeezing is directional in phase space: it reduces noise in one direction and increases it in the other. If the measurement is inadvertently rotated relative to the squeezed quadrature, the experimenter ends up measuring in the anti-squeezed direction, making the result worse than that of standard laser light. Maintaining phase lock between the squeezed source and the local oscillator of the homodyne detector is therefore an essential engineering requirement, and active feedback control systems are typically used to stabilize this phase relationship in real experiments.

Diagram 6.8 - Shot-Noise vs. Squeezed-Noise Oscilloscope Traces

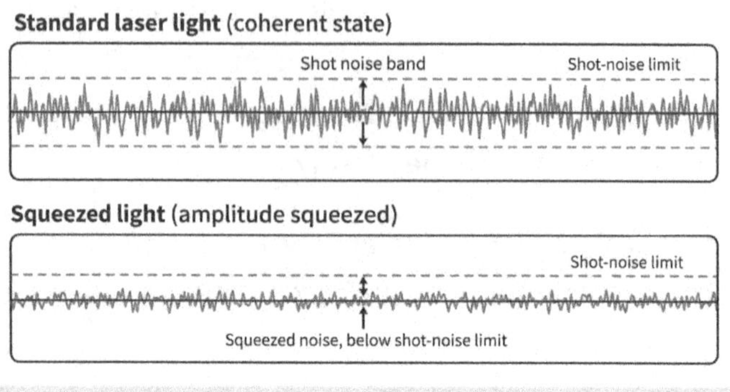

Squeezed light measurement traces show visibly reduced noise compared to standard laser measurements.

LIGO and the Quantum Limits of Gravitational Wave Detection

To appreciate the scale of what LIGO achieves, it helps to have a concrete sense of the measurement it is performing. Gravitational waves are ripples in the fabric of spacetime produced by accelerating masses. When two black holes spiral inward and merge, they release a tremendous burst of gravitational energy. Yet by the time those gravitational waves reach Earth after traveling for hundreds of millions or billions of light-years, the resulting strain, the fractional change in length they produce, is roughly one part in ten to the twenty-first power. Applied across LIGO's four-kilometer arms, this corresponds to a physical movement of the mirrors of about one-thousandth the diameter of a proton.

LIGO is a Michelson interferometer on a grand scale. A laser beam enters and is split into two perpendicular arms, each four kilometers long, terminated by high-reflectivity mirrors. The light bounces back and forth hundreds of times before being recombined. In the absence of a gravitational wave, the two arms are tuned so that the recombined beams cancel completely at the detector, producing no signal. When a gravitational wave passes, one arm is briefly stretched while the other is compressed, changing the relative length of the arms by a tiny amount. This breaks the cancellation, sending a small signal to the detector whose shape and amplitude encode the passing wave.

The displacement LIGO must detect corresponds to roughly one-thousandth of a proton's diameter over a distance of four kilometers. At such extremes, quantum shot noise in the laser light is a fundamental limiting factor at the high-frequency end of the detection band, where signals from binary black hole mergers are strongest. The photons arriving at the detector from the interferometer are not perfectly steady; their random arrival times create shot noise that masks the gravitational-wave signal. More photons reduce this noise, but increasing laser power introduces a different quantum effect, radiation-pressure noise, at low frequencies. The two quantum noise contributions place competing demands on the design, and the only way to reduce both simultaneously is to use quantum-engineered light.

Diagram 6.9 - LIGO Interferometer with Squeezed-Light Injection

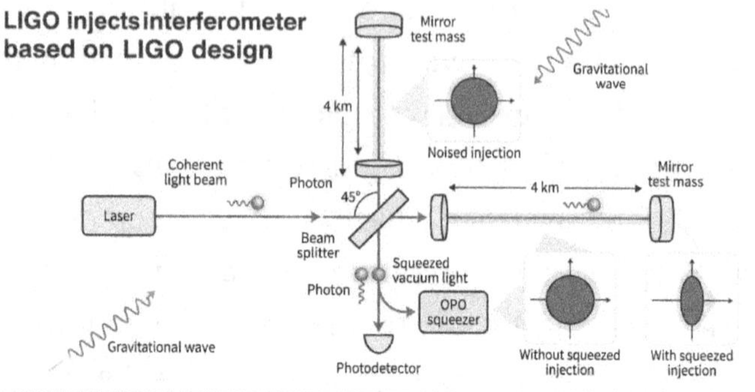

LIGO injects squeezed vacuum into its output port to reduce shot noise and push sensitivity beyond the standard quantum limit.

The solution, implemented at LIGO's third observing run in 2019, is to inject squeezed vacuum into the interferometer's output port. Vacuum fluctuations normally enter through that port and add shot noise to every measurement. Replacing them with a squeezed vacuum state, a state with reduced noise in the relevant quadrature, lowers the shot-noise floor without changing the rest of the interferometer. The squeezing must be phase-aligned with the measured quadrature, an engineering challenge that has been successfully solved.

The improvement has been roughly 3 dB in sensitivity at high frequencies, equivalent to about a factor of 1.4 in amplitude noise. This translates to a significantly larger detectable volume of the universe because gravitational-wave strain falls off with distance: extending the detectable range by 1.4 times the distance increases the detectable volume by roughly a factor of 2.8. The deployment of squeezed light at LIGO is the most prominent operational demonstration of quantum-enhanced sensing in the world, marking the transition of quantum optics from a laboratory science into a precision engineering tool. Future detectors, including the proposed Einstein Telescope and Cosmic Explorer, are being designed with quantum noise management as a primary engineering requirement from the outset.

Balanced Homodyne Detection: Listening to Quantum States

To benefit from squeezed light, you need not only a source but also a measurement technique that accesses the squeezed quadrature specifically. Standard photodetectors measure intensity, which corresponds to the time-averaged square of the electric field amplitude. They do not directly reveal the quadrature structure of the quantum state. To measure a specific quadrature with quantum-limited precision, physicists use a technique called balanced homodyne detection.

In balanced homodyne detection, the signal beam is combined on a 50/50 beam splitter with a strong reference beam called the local oscillator, a bright coherent laser beam at the same frequency as the signal, with a precisely controlled phase. The 50/50 beam splitter produces two output beams, each containing a mixture of signal and local oscillator. Two photodetectors, one for each output, measure the intensity of these mixed beams. The key step is subtracting the two photocurrents. Subtraction cancels the local oscillator's own shot noise, which would otherwise overwhelm the signal, because the 50/50 beam splitter distributes the local oscillator equally to both outputs. What remains after subtraction is a signal proportional to one quadrature of the input beam, specifically the quadrature aligned with the phase of the local oscillator.

Diagram 6.10 - Balanced Homodyne Detection Setup

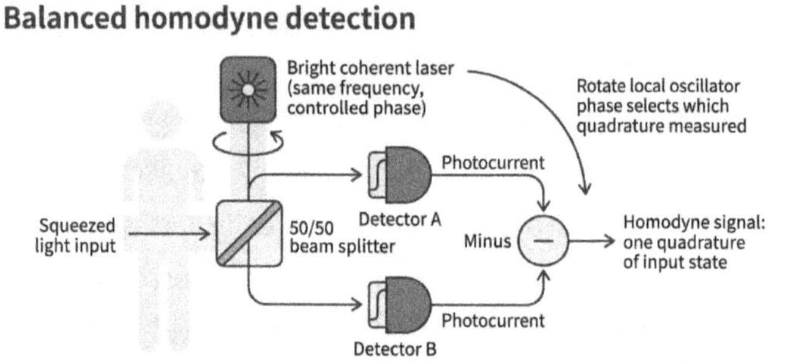

Balanced homodyne detection

Balanced homodyne detection isolates a single quadrature by subtracting two detector signals, canceling local oscillator noise

By rotating the phase of the local oscillator, the experimenter selects which quadrature to measure: one orientation gives the X quadrature, a ninety-degree rotation gives the P quadrature. When the input state is squeezed, the homodyne measurement shows reduced noise when the local oscillator is aligned with the squeezed quadrature, and increased noise when aligned with the anti-squeezed quadrature. Plotting noise level versus local oscillator phase traces out the shape of the quantum state in phase space, a process called quantum state tomography, and the standard diagnostic method for verifying and characterizing squeezed-light sources.

Balanced homodyne detection is also the foundation for continuous-variable quantum information processing. This approach encodes quantum information in the continuous quadrature values of a light field rather than in discrete photon states. Systems built around homodyne detection can operate at very high bandwidths and near-unit detection efficiency using standard silicon photodiodes, a practical advantage over the cryogenic single-photon detectors required for photon-counting-based quantum information schemes. Understanding the homodyne technique is therefore relevant not only to precision measurement but to a growing range of quantum computing and quantum communication architectures.

Quantum Noise as Both a Limit and a Resource

For most of the twentieth century, quantum noise was treated as a nuisance: an irreducible background that set the floor of what was measurable, designed around rather than engaged with. Shot noise was a budget line in a sensitivity calculation, something to minimize by using more photons, but fundamentally fixed in character. The quantum mechanical nature of light was, from this perspective, a limitation on human ingenuity.

The rise of quantum optics has completely inverted this perspective. Quantum noise is not merely a limit; it is a physical resource with rich structure that can be shaped, engineered, and exploited. The uncertainty principle is the engine behind squeezed-state generation: squeezing redistributes uncertainty. The vacuum fluctuations that feed shot noise are also the fluctuations that seed spontaneous emission and enable the parametric processes that produce entangled photon pairs. The randomness of individual photon arrivals, which creates shot noise in classical measurements, is precisely the randomness that makes quantum key distribution secure: an eavesdropper cannot intercept photons without introducing detectable perturbations, because the quantum states are fundamentally probabilistic.

This shift has direct engineering consequences. Classical sensors improve sensitivity by increasing photon count, which reduces shot noise by a factor of the square root of the photon number. Quantum-enhanced sensors using squeezed or entangled light can approach Heisenberg-limited sensitivity, in which the improvement scales with the photon number itself rather than its square root. In practice, loss and decoherence limit how close real experiments get to this limit, but even a modest improvement represents a qualitative change in capability. Quantum imaging exploits the same logic: photon correlations enable better images with fewer photons, a valuable trade-off in biological imaging, where excessive light damages living cells.

Diagram 6.11 - Quantum Noise as Resource vs. Limit

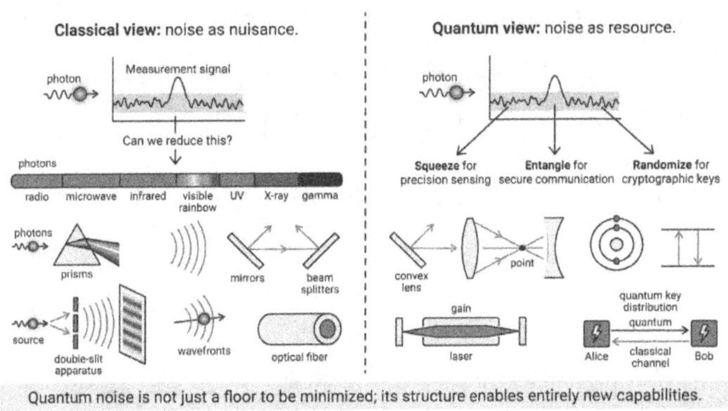

Quantum noise is not just a floor to be minimized; its structure enables entirely new capabilities.

Quantum noise is also the conceptual bridge to the chapters that follow. The quantum correlations between photon pairs produced by parametric down-conversion, the same process that generates squeezed light, are the physical substrate for quantum entanglement. The quantum randomness that makes single-photon states unpredictable is the randomness that makes quantum cryptographic keys unguessable. The Heisenberg limits on measurement precision are the same limits that guarantee the security of quantum communication protocols. Everything in the quantum optics landscape connects through the structure of the quantum state.

This interconnection means that the concepts introduced in this chapter are not isolated topics; they are the foundation on which the remaining chapters build. A reader who understands coherent states, Fock states, and squeezed states, and who has internalized the phase-space picture, will find that entanglement, quantum key distribution, and photonic quantum computing all become substantially more intuitive. The quantum state is the thread that ties quantum optics together as a unified discipline, and the noise landscape of this chapter is the terrain on which everything else is built.

Diagram 6.12 - Quantum States of Light: A Conceptual Map

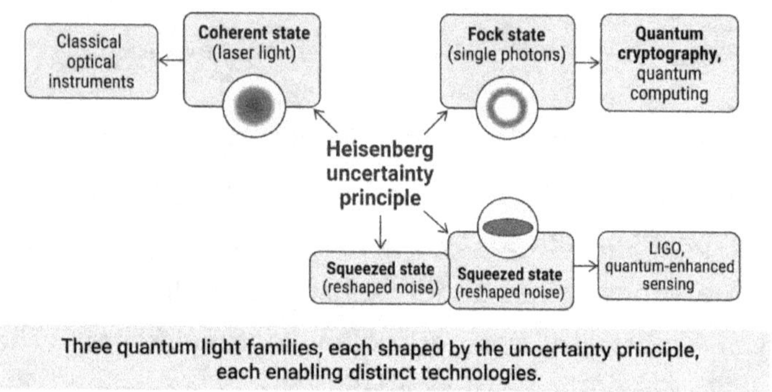

Three quantum light families, each shaped by the uncertainty principle, each enabling distinct technologies.

What This Means For You

If you work in precision instrumentation, defense sensing, telecommunications, or scientific research, the concepts in this chapter have immediate practical relevance. The shot-noise limit is not a distant theoretical abstraction; it is the noise floor encountered in any precision optical measurement, from laser ranging and LIDAR to optical gyroscopes and atomic clocks. Knowing where the shot-noise floor lies, how it scales with photon number, and how squeezed light can surpass it gives you a framework for evaluating a sensing system's fundamental limits and the quantum engineering needed to exceed them.

If your work involves quantum technology development, understanding the distinction between coherent states, Fock states, and squeezed states is essential for reading the literature and evaluating experimental platforms. A device producing 'squeezed vacuum at 15 dB below shot noise' is making a specific claim about its output quantum state. A source producing 'indistinguishable single photons at 98% purity' is referring to the Fock-state character of its output, and the engineering significance of that purity number becomes clear only when you understand the difference between Poisson photon statistics and the fixed photon count of a true single-photon state. This vocabulary gives you the tools to evaluate claims substantively rather than taking them at face value.

For technology strategists and policy professionals, the key takeaway is that quantum optics is not a monolithic field but a collection of distinct engineering capabilities at different maturity levels. Squeezed-light technology is the most mature, already deployed at LIGO and moving toward commercial sensing applications. Single-photon sources and photon-number-resolving detectors are at the next tier, moving from research into early commercial products for quantum key distribution and quantum computing. Understanding the conceptual differences between these technology families enables more informed assessments of what is ready for deployment, what is still at the research stage, and which questions to ask when evaluating a vendor's claims or a program's technical approach.

Takeaway

Quantum states of light are not a single thing but a rich family of configurations, each characterized by a different pattern of quantum uncertainty in phase space. The coherent state, produced by an ideal laser, distributes its minimum uncertainty symmetrically between two quadratures and behaves as classically as quantum mechanics allows. The Fock state has a precisely defined photon number and a completely undefined phase, the starkest possible departure from classical wave behavior. The squeezed state redistributes quantum noise so that one quadrature has less uncertainty than the shot-noise limit, at the cost of increased uncertainty in the complementary quadrature. Each of these states is a real, physically producible, experimentally verifiable configuration of the electromagnetic field.

Shot noise, the irreducible fluctuation from the discrete arrival of individual photons, sets the standard quantum limit for measurements made with coherent laser light. This limit cannot be overcome by classical engineering; it can only be surpassed by quantum engineering of the light state itself. Squeezed light accomplishes exactly this. Its most prominent operational use is at LIGO, where it has extended the sensitivity of the world's most precise measuring instrument beyond the shot-noise floor, enabling the detection of gravitational waves from merging black holes across cosmological distances. Balanced homodyne detection, which combines a signal beam with a phase-controlled local oscillator and subtracts the two detector outputs, is the standard technique for accessing and verifying the quadrature-noise properties of squeezed states.

The broader conceptual lesson is the shift from viewing quantum noise as a pure constraint to recognizing it as a physical resource with exploitable structure. The same Heisenberg uncertainty that creates the shot-noise floor also enables the squeezing that breaks it. The same quantum correlations that add noise in one measurement basis create security in another. The same randomness that limits one kind of sensing powers the unpredictability required by quantum cryptography. This is the signature of a physical framework richer than the classical physics it superseded, and the remaining chapters build directly on it: quantum entanglement, unbreakable communication, and computers made of light all follow from the quantum state landscape described here.

The noise-canceling headphones at the start of this chapter illustrated a classical noise floor that cannot be escaped within the classical paradigm. Squeezed light demonstrates that the quantum noise floor, equally immovable in appearance, is in fact a reshapeable landscape. Moving through that landscape requires quantum engineering, the deliberate manipulation of quantum states. This chapter has described the terrain and the tools. The chapters ahead describe what you can build once you know how to navigate them.

7 Entanglement: When Photons Share Secrets

Opening Scenario

Imagine two sealed envelopes, each containing a single coin. You prepare both envelopes at the same table, seal them without looking inside, and mail one to a friend on the West Coast and the other to a friend on the East Coast. You ask both friends to open their envelopes at exactly noon on the same day and to record whether their coin is heads or tails. When they compare notes afterward, you discover something striking: every single time one friend found heads, the other found tails. Without exception. Every trial. Three thousand miles apart, neither knew what the other would find, and yet the results were perfectly, consistently opposite.

The natural classical explanation is reassuring: the coins were preset. Before you sealed the envelopes, one coin was already heads-down, and the other was already tails-down. The result was determined at the time of packing, not at the time of opening. The coins carried hidden instructions, and your friends read them when they opened their envelopes. This is the most intuitive account, and for three centuries of classical physics, it would have been the only account anyone seriously considered.

Quantum mechanics says something entirely different. It says the coins did not have fixed values before they were opened. The act of opening one envelope decided the result of both, simultaneously, regardless of the distance between them. More astonishing still: this is not a philosophical position. It is a proven experimental fact. Physicists have tested this question using real photons, real detectors, and real distances, and the experiments have returned a clear verdict. The quantum account is correct, and this chapter explains how we know that, what it means, and what it makes possible.

Diagram 7.1 - The Entangled Coin Thought Experiment

Executive technical illustration

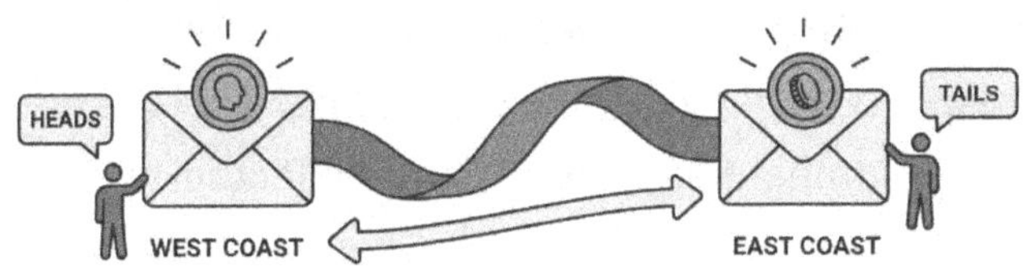

Perfectly correlated results, no matter the distance, with no signal between them.

Why It Matters

Quantum entanglement is not a curiosity locked inside university laboratories. It is the enabling foundation of three distinct and growing technology sectors. Quantum cryptography uses entangled photon pairs to generate encryption keys that are provably unbreakable by the laws of physics. Quantum teleportation uses entanglement to transfer the complete quantum state of a photon to a distant location without physically moving the photon.

Quantum computing exploits entanglement between photonic qubits to perform calculations that no classical machine could replicate. In each case, the same strange correlation that puzzled Einstein is doing practical work.

Beyond technology, the experimental proof of entanglement resolved one of the deepest debates in the history of science. Albert Einstein spent the last decades of his life insisting that quantum mechanics was incomplete: surely, he argued, particles must carry hidden information about their outcomes before measurement. The experimental work that earned John Clauser, Alain Aspect, and Anton Zeilinger the 2022 Nobel Prize in Physics settled that argument definitively, in the opposite direction from Einstein's intuition. Entanglement is real, verified, and increasingly commercial. Understanding it is now a professional competency, not just an intellectual curiosity.

What Entanglement Means in Plain Language

To understand entanglement, you first need a firm grasp of what quantum mechanics says about measurement. In classical physics, a coin in a sealed envelope has a definite state, heads or tails, whether or not anyone is looking. The envelope merely hides information that already exists. Quantum mechanics says something more radical: a quantum system such as a photon genuinely does not have a definite value for certain properties, including polarization, until a measurement is made. Before measurement, the photon exists in a superposition, carrying a weighted possibility for each outcome, but neither outcome is real yet. The measurement does not merely reveal a pre-existing fact. It creates the fact.

Polarization refers to the orientation in which a photon's electric field oscillates. Think of a jump rope being shaken: the wave can travel in a vertical plane, a horizontal plane, or at any angle. A photon in a superposition of vertical and horizontal has no definite answer until a polarizer forces one. This is not ignorance about the photon's true state. According to quantum mechanics, as confirmed by experiment, there is no truer state than ignorance.

Now imagine two photons that are entangled. They share a single joint quantum description called a quantum state that cannot be broken down into two independent descriptions. Neither photon has a definite polarization on its own. But their joint state encodes a strict correlation: if you measure one photon as vertical, the other will be measured as horizontal. This holds regardless of the distance between the photons and regardless of how quickly the measurements are made. It holds even when the choice of measurement direction is made randomly, at the last microsecond before detection, with no time for any signal to travel between the two sites.

This is what physicists mean when they call entanglement non-local. The word local refers to the classical principle that a particle can only be influenced by things physically adjacent to it. Since information cannot travel faster than light, a measurement in Los Angeles should not instantly affect an experiment in New York if locality holds. Entanglement appears to challenge this, and a key goal of this chapter is to clarify exactly what that appearance does and does not imply.

Diagram 7.2 - Entangled Photon Pair Visualization

ENTANGLED STATE

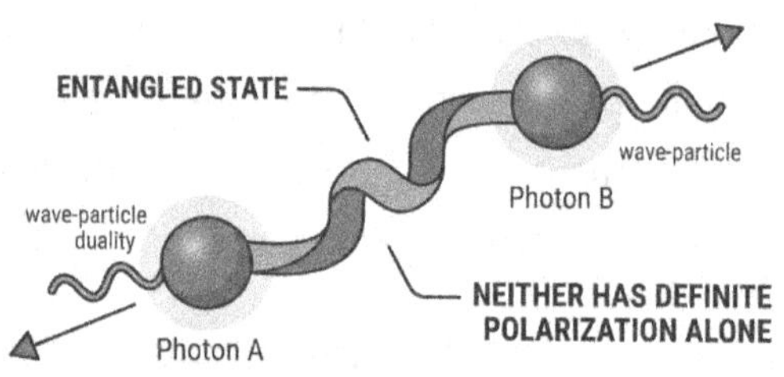

An entangled pair shares one quantum description; neither photon has a definite state until measurement.

A Bell state, also called an EPR pair in honor of Einstein, Podolsky, and Rosen, who first theorized such correlations in 1935, is a photon pair in which entanglement is as strong as quantum mechanics permits. In one Bell state, the two photons always give opposite polarization results; in another, they always give the same result. All four Bell states have been created and studied in laboratories worldwide.

Entanglement is not the same as simple correlation. Two coins packaged to be opposite are classically correlated: they carry hidden instructions. The correlation between entangled photons is stronger than any hidden instructions could produce. This excess above the classical limit is what Bell's theorem quantifies, and the Nobel Prize experiments verified. It is the entire point and the source of entanglement's power as a practical resource.

How Entangled Photon Pairs Are Created

The most widely used method for producing entangled photon pairs is called spontaneous parametric down-conversion, typically abbreviated as SPDC. The name sounds formidable, but the idea is straightforward. Parametric refers to exploiting the nonlinear optical properties of a specially engineered crystal. Down-conversion means a single higher-energy photon is converted into two lower-energy photons. Spontaneous means the process occurs at unpredictable moments, without being triggered by anything specific, governed solely by quantum probability.

In practice, a pump laser fires photons at a nonlinear crystal. Occasionally, with a probability of roughly one in a billion photon-crystal interactions, a pump photon is absorbed, and two new photons emerge. These two daughter photons share the energy and momentum of the original pump photon. Because they are born from the same parent in a single quantum event, they share entangled quantum states from the moment of their creation. Their polarizations are correlated in a way that exceeds any classical pre-arrangement.

Diagram 7.3 - Spontaneous Parametric Down-Conversion Crystal

Spontaneous Parametric Down-Conversion

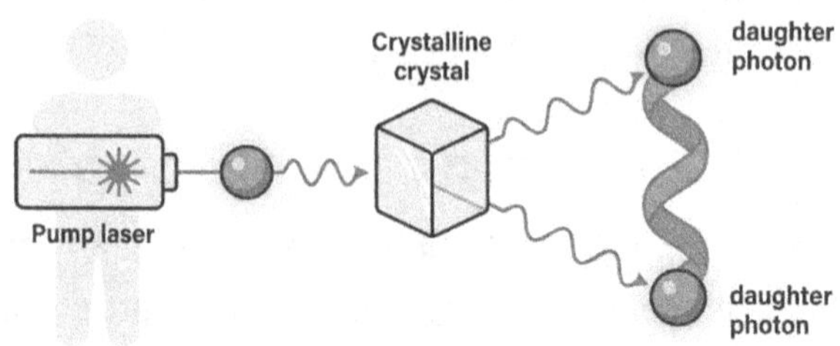

One pump photon enters; two entangled daughter photons emerge at opposite angles.

The probability of one in a billion sounds discouraging, but a milliwatt pump laser fires billions of photons per second, so that an SPDC source can produce millions of entangled pairs per second. Pair production is heralded: the detection of one photon in the pair can signal the existence of its partner, allowing downstream systems to prepare for an incoming entangled photon rather than waiting unthinkingly. This heralding property makes SPDC sources practical for real quantum communication links and laboratory Bell tests.

SPDC is not the only method. Atomic cascade emission, in which an excited atom emits two photons in rapid succession through a cascade of energy levels, was used in the earliest Bell test experiments, including John Clauser's pioneering work and Alain Aspect's landmark 1982 experiments. Quantum dot sources, tiny semiconductor structures that can be engineered to emit entangled photon pairs on demand, have recently emerged as promising platforms for practical quantum networks. Once entangled pairs are produced, they are guided into optical fiber or free-space paths and routed to two separate detectors. The entanglement persists as long as the photons avoid decoherence, and advances in low-loss fiber have enabled the practical distribution of entangled photons over hundreds of kilometers.

Diagram 7.4 - Entangled Photon Distribution to Alice and Bob

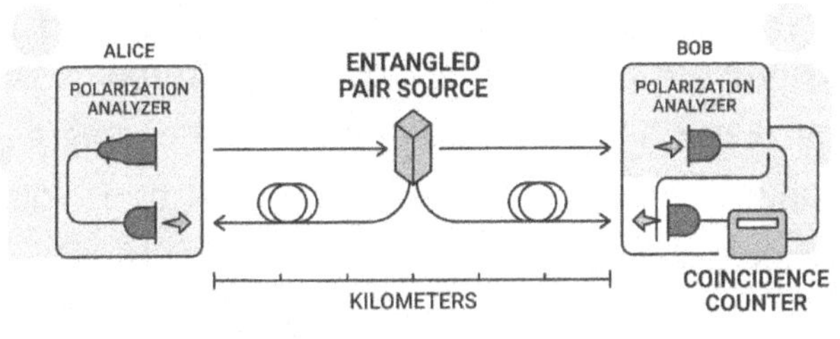

Alice and Bob each receive one photon of an entangled pair, separated by kilometers of optical fiber.

Bell's Theorem: The Most Profound Result in Quantum Foundations

In 1964, the Northern Irish physicist John Stewart Bell published a paper now recognized as one of the most important theoretical results of the twentieth century. Bell was grappling with the question Einstein had raised decades earlier: Is quantum mechanics complete? Einstein believed particles must carry definite properties before measurement, properties that quantum mechanics fails to describe. This idea is called a hidden-variable theory, because the decisive information is hidden from quantum mechanics but presumably exists in physical reality.

Bell asked: Is there an experiment that could distinguish between a world with hidden variables and a world where quantum mechanics is truly complete? He found that there is. He derived a mathematical inequality, now called Bell's inequality, that any theory based on local hidden variables must obey. Local realism means two things: physical effects are local, so a distant measurement cannot instantly affect a local particle, and particles carry real, definite properties before measurement. Bell's inequality sets an upper bound on the degree of correlation between the results of two distant measurements, given these assumptions.

The crucial insight was that quantum mechanics predicts correlations that violate Bell's inequality. If you measure the polarizations of two entangled photons at various angles, the correlation follows a cosine-squared relationship with the angle between the measurement settings. For certain angle combinations, this quantum prediction exceeds the maximum value permitted by any local hidden-variable theory. The excess is large enough to be clearly detectable in laboratory experiments. Bell transformed the debate from philosophy into experiment.

Diagram 7.5 - Polarization Correlation Curve and Bell Inequality

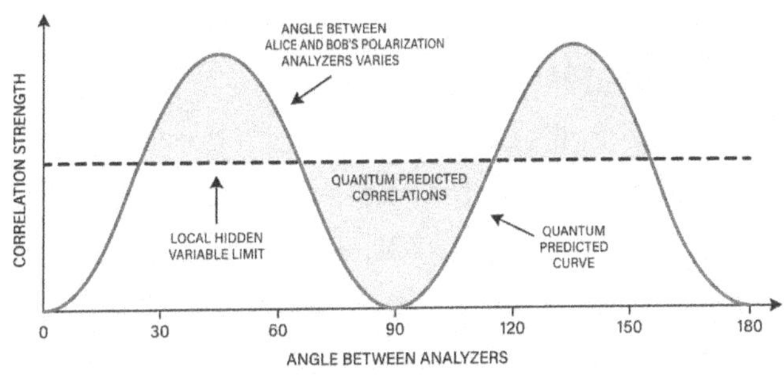

Quantum mechanics predicts correlations that exceed the limit any classical hidden-variable theory could produce.

Bell's theorem is sometimes called the most profound result in quantum foundations because it converted a philosophical argument into an empirical question. The prediction was clean: local hidden variables require correlations below Bell's bound; quantum mechanics requires correlations above it. Nature had a vote, and the experiments went on to cast it decisively.

The most commonly tested version of Bell's inequality, called the CHSH inequality after Clauser, Horne, Shimony, and Holt, involves four combinations of measurement angles between Alice and Bob. For any local hidden-variable theory, a specific sum of correlation values cannot exceed two. Quantum mechanics predicts a maximum of approximately 2.83. Every loophole-free Bell test experiment conducted to date has found values significantly above two, matching the quantum prediction and definitively ruling out local hidden variables.

The Aspect and Zeilinger Experiments and the 2022 Nobel Prize

The story of how Bell's theorem was tested experimentally is one of the great scientific detective narratives of the modern era. In the early 1970s, John Clauser performed the first serious test using entangled photons from atomic cascade emission. His results favored the quantum prediction, but the experiment had loopholes. The detection loophole arose because detectors captured only a fraction of photons: if the detected sample were biased, a hidden-variable theory might explain the correlations. The locality loophole arose because measurement settings were chosen before photons were emitted, leaving open the possibility that photons carried information about upcoming settings in their hidden variables.

Diagram 7.6 - EPR Debate and Einstein's Spooky Action

Einstein believed entangled particles carried hidden instructions.
Experiment proved him wrong.

Alain Aspect, at the Institut d'Optique in Orsay, France, addressed the locality loophole in a pivotal set of experiments completed in 1982. Aspect's key innovation was to rapidly switch the measurement settings on each side, using acousto-optical modulators that changed the polarization angle while the photons were already in flight. By switching faster than the time it takes light to travel between the two detectors, Aspect ensured that Alice's choice could not have influenced Bob's photon via any light-speed signal. His results again violated Bell's inequality, this time with the locality loophole substantially closed.

Anton Zeilinger, at the University of Vienna and later the Austrian Academy of Sciences, pushed experimental entanglement to progressively greater distances, from fiber loops across Vienna to free-space links between the Canary Islands spanning 144 kilometers. His group demonstrated quantum teleportation experimentally in 1997 and generated three-photon entangled states called GHZ states, named for Greenberger, Horne, and Zeilinger, which provide a sharper test of quantum foundations than two-photon Bell tests. The definitive closure of both loopholes simultaneously came in 2015, when three independent groups published loophole-free Bell test results within months of each other. The experiment led by Ronald Hanson in Delft was particularly rigorous, and these 2015 results closed the last serious scientific debate: quantum entanglement is a genuine feature of nature.

In October 2022, the Royal Swedish Academy of Sciences awarded the Nobel Prize in Physics jointly to John Clauser,

Alain Aspect, and Anton Zeilinger for their experiments with entangled photons, which established the violation of Bell inequalities and pioneered quantum information science. The Nobel citation specifically noted the implications for quantum key distribution, quantum teleportation, and quantum computing. It was a rare Nobel Prize that recognized not merely a discovery but an experimental proof about the fundamental nature of physical reality.

Diagram 7.7 - Bell Test Setup with Random Setting Choices

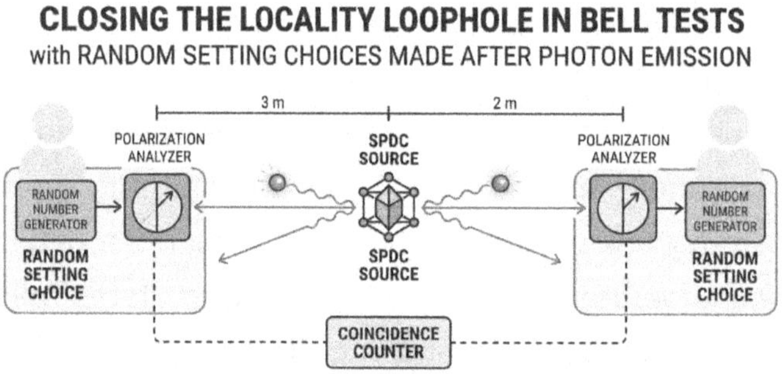

Random setting choices made after photon emission close the locality loophole in Bell tests.

Diagram 7.8 - 2022 Nobel Prize in Physics Recognition

Clauser, Aspect, and Zeilinger proved that quantum entanglement is real and violates Bell inequalities.

Why Entanglement Cannot Be Used for Faster-Than-Light Signaling

One of the most common questions people ask when they first encounter entanglement is: if measuring one photon instantly determines the state of the other, regardless of distance, why can't that be used to send information faster than light? It is a natural question, and the answer reveals something deep about quantum mechanics and the structure of spacetime.

The key is that Alice cannot control the outcome of her measurement. When she measures her photon's polarization, the result is random. Quantum mechanics is irreducibly probabilistic: the outcome is genuinely undetermined before measurement, and Alice has no lever to force a specific result. If she could choose her outcome, she could encode a message into Bob's photon's state. But she cannot choose. She only witnesses.

Bob is in the same position. He measures his photon and gets a random result. Nothing in his outcome tells him whether Alice has measured yet, what setting she used, or what she got. His sequence of results looks like a random string of ones and zeros, indistinguishable from noise. No message is present.

The correlation between Alice's and Bob's results is real and quantum-mechanical. Still, it is only visible after the fact, when they compare their records through an ordinary classical communication channel: a phone call, an email, or a data link that travels at or below the speed of light. That classical comparison is the limiting step. The no-communication theorem of quantum mechanics formally guarantees that entanglement alone cannot transmit information. Faster-than-light communication remains impossible, and quantum mechanics carefully preserves that constraint even while producing correlations that feel non-local.

Entanglement is non-local in the sense that local hidden variables cannot explain correlations. It is not non-local in the sense that it allows anyone to send a message. Quantum mechanics maintains this distinction precisely. Einstein's worry that entanglement undermined special relativity was unfounded, though his discomfort pointed to something genuinely profound about the structure of nature.

Diagram 7.9 - Why Entanglement Cannot Signal Faster Than Light

16:9 INFOGRAPHIC DIAGRAM

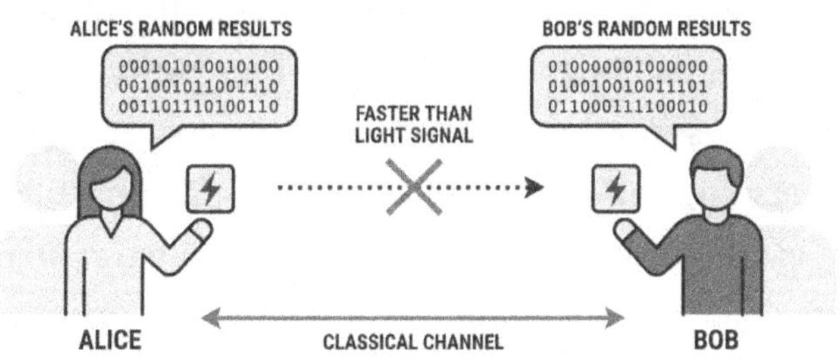

Each side sees only random results; the correlation only becomes visible after a classical comparison.

Entanglement as a Resource for Quantum Information

If entanglement cannot transmit information on its own, why is it called a resource? Entanglement, combined with classical communication, enables capabilities that are impossible without it. The combination is strictly more powerful than classical communication alone, even though neither ingredient by itself breaks any physical law. This pattern runs throughout quantum information science: quantum resources expand what is achievable not by circumventing physical constraints but by exploiting correlations that have no classical analog.

The most dramatic example is quantum teleportation. Teleportation in the quantum sense does not mean transporting matter. It means transferring the complete quantum state of one photon to another photon at a distant location, without physically moving the original photon. Alice has a photon whose quantum state she wants to transfer to Bob. Alice and Bob also share a pre-distributed entangled pair. Alice performs a joint measurement on her photon and her half of the entangled pair, then sends Bob two classical bits describing the result. Bob uses those two bits to apply a specific operation to his half of the entangled pair, and his photon is then in the same quantum state as Alice's original photon. The original state is destroyed in Alice's measurement, consistent with the no-cloning theorem. The state has been teleported, not copied.

Diagram 7.10 - Quantum Teleportation Circuit

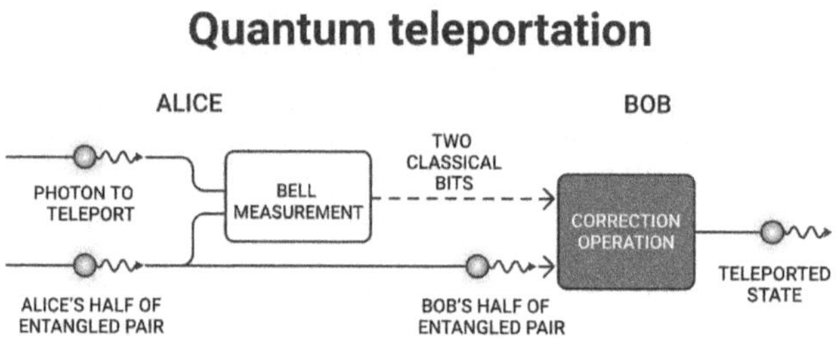

Quantum teleportation moves a photon state to a distant location using entanglement and two classical bits.

Quantum teleportation was first demonstrated experimentally by Zeilinger's group in 1997 and has since been achieved over fiber and free-space links spanning dozens of kilometers, with the Chinese Micius satellite reaching distances above 1,400 kilometers. In quantum networking, teleportation allows quantum states to be routed through a network without directly transmitting the physical photon, which is valuable when direct transmission over long distances is lossy.

Entanglement is also the key ingredient in entanglement-based quantum key distribution. In this protocol, Alice and Bob each receive one photon of an entangled pair and measure in randomly chosen bases. When they compare their basis choices over a classical channel, keeping only results where they chose the same basis, they obtain a shared string of random bits. Any eavesdropper who tries to intercept the photons disturbs the entanglement, reducing the measured correlations below the Bell inequality violation threshold and alerting Alice and Bob to the intrusion. This approach, called the E91 protocol after Artur Ekert, who proposed it in 1991, uses entanglement as its physical security guarantee.

In photonic quantum computing, entanglement between photonic qubits is the resource that enables quantum logic gates. A photonic qubit is a quantum bit encoded in the polarization or path of a photon. Photonic qubits can be entangled through linear optical elements, beam splitters, and coincidence measurements in a scheme known as the KLM protocol, after Knill, Laflamme, and Milburn, who proved in 2001 that universal quantum computation with photons is achievable using linear optics and photon counting alone. Modern photonic quantum processors realize entanglement between on-chip qubits using silicon photonic waveguides that route photons through interferometers with subnanometer precision. Entanglement is to quantum computing what electrical current is to a classical processor: the fundamental carrier of the computational work.

Diagram 7.11 - GHZ Three-Photon Entangled State

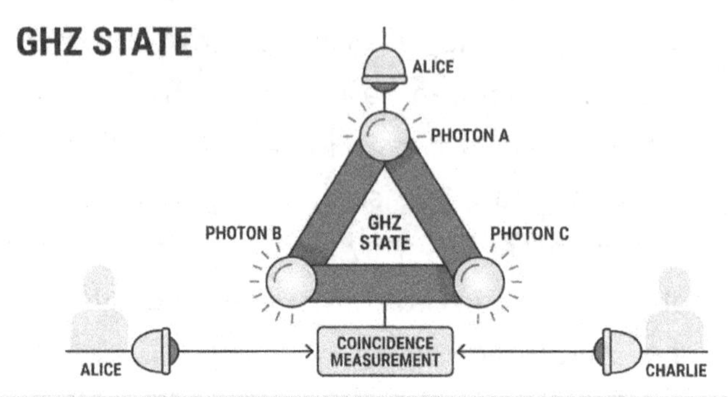

A GHZ state entangles three photons simultaneously and provides an even sharper test of quantum reality.

The Monogamy of Entanglement and Its Security Implications

Among the properties of entanglement that matter most for practical applications, one stands out for its elegance and its consequences: the monogamy of entanglement. Monogamy is the formal quantum mechanical constraint that a photon maximally entangled with one partner cannot simultaneously be entangled with any other particle. Entanglement, unlike classical correlations, is exclusive. The more strongly photon A is entangled with photon B, the less it can be entangled with photon C or anything else.

Consider what an eavesdropper, conventionally called Eve, would need to do to intercept an entangled quantum key distribution link without detection. Eve must intercept one photon from each pair, extract information from it, and forward a photon to Bob that appears to be the original entangled partner. But extracting information requires measuring or interacting with the photon, which in turn entangles it with Eve's detector. Once Eve's system is entangled with Alice's photon, the monogamy constraint guarantees that Alice's photon is less entangled with Bob's photon. When Alice and Bob check their correlations, the reduction in entanglement shows up as a deviation from the expected Bell inequality violation. Eve is detected, not by any technological countermeasure, but by a fundamental law of physics.

This is fundamentally different from classical cryptographic security. Classical encryption security rests on computational hardness: breaking it requires solving a mathematical problem that is currently too slow to complete in practice. Computational hardness is not a law of physics; a powerful enough computer could, in principle, break it. Entanglement-based security does not rely on computational hardness at all. It relies on the monogamy of entanglement, a consequence of the mathematical structure of quantum state spaces. No computer, classical or quantum, can break monogamy. It is not an algorithm. It is physics.

The Hong-Ou-Mandel effect, named for the three physicists who discovered it in 1987, is a two-photon interference phenomenon closely connected to entanglement and now an indispensable diagnostic tool in quantum optics. The Hong-Ou-Mandel dip appears when two truly identical photons are directed into the two input ports of a beam splitter simultaneously. Classical intuition says each photon should have a fifty-fifty chance of being reflected or transmitted, so sometimes both should exit from the same port and sometimes from opposite ports. But if the photons are truly identical in every degree of freedom, frequency, polarization, and timing, they always exit from the same port. The coincidence rate between two detectors monitoring the output ports drops to exactly zero. This perfect bunching is a quantum interference effect with no classical analog.

Diagram 7.12 - Hong-Ou-Mandel Dip

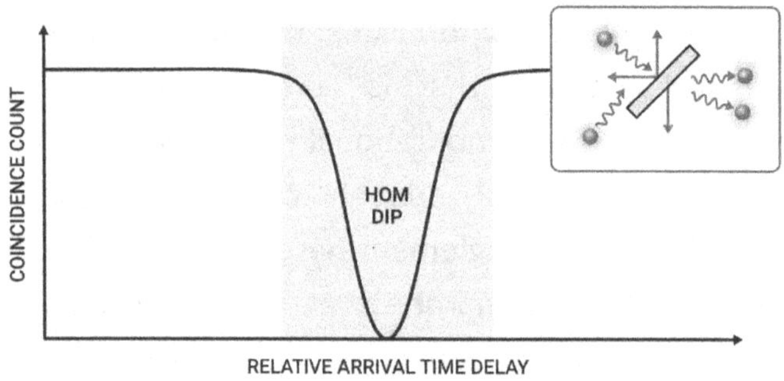

Identical photons always bunch at a beam splitter; the HOM dip confirms their indistinguishability.

The Hong-Ou-Mandel dip is the standard test for photon indistinguishability, the degree to which two photons are truly identical. Indistinguishability is a prerequisite for many quantum information protocols, including the linear optical gates used in photonic quantum computing. A high-visibility dip, meaning one that drops close to zero, indicates photons suitable for quantum information processing. Poor visibility indicates distinguishable photons that will not produce useful quantum interference. Engineers building quantum photonic systems routinely measure HOM visibility as a quality metric for their photon sources.

Monogamy also shapes the architecture of quantum repeaters, devices that extend the reach of entanglement-based networks across successive segments. A repeater cannot amplify a quantum signal, which would violate the no-cloning theorem; instead, it creates fresh entangled pairs at each segment and performs entanglement swapping to connect them into a single end-to-end link. The monogamy constraint determines how many simultaneous users a repeater can serve, an engineering parameter with direct implications for future quantum internet design.

What This Means For You

If you work in cybersecurity, network engineering, or telecommunications, entanglement is already a practical concern. Governments and research consortia in the United States, Europe, and China are building entanglement-based quantum network testbeds. Within a decade, quantum key distribution links using entangled photon pairs are expected to become commercially available. Understanding how entanglement-based security differs from classical cryptographic security is a genuine professional competency for anyone involved in long-term network security planning.

If you work in hardware development or photonics engineering, the production and detection of entangled photon pairs are rapidly growing product areas. SPDC sources, quantum dot emitters, and superconducting nanowire single-photon detectors (SNSPDs) are moving from research instruments to manufacturable components. Companies including ID Quantique, Quandela, and PsiQuantum are commercializing entanglement-based hardware. Engineers who understand SPDC, coincidence counting, and the HOM effect are well-positioned to contribute.

If your interest is quantum computing, entanglement is foundational. Every proposed architecture for a universal quantum computer, whether photonic, superconducting, or trapped ion, relies on entanglement for qubit interactions and quantum logic. Without it, a quantum computer offers no advantage over a classical probabilistic machine. With it, computational resources scale exponentially, enabling problems in optimization, simulation, and cryptanalysis to be solved at speeds no classical machine could approach.

Takeaway

Two photons, born from the same quantum event, share a bond that any classical picture of reality cannot explain. The bond is not a hidden message written before they parted. It is not a signal traveling between them. It is a genuine quantum correlation, proven by experiment to exceed every limit that local hidden-variable physics would allow, and certified by the Nobel Prize in Physics as one of the defining discoveries of the era. The coins were not preset. The measurement decided everything.

The phenomenon began as Einstein's discomfort, became Bell's theorem, was tested by Clauser's pioneering apparatus, was sharpened by Aspect's switching experiments, was pushed to continental and satellite distances by Zeilinger's group, and was definitively settled in the loophole-free tests of 2015. At every step, the quantum prediction held. At every step, the classical alternative was ruled out. The universe is stranger than Einstein could accept, but the strangeness is real, measurable, and increasingly useful.

Entanglement is a resource. Photon pairs produced by spontaneous parametric down-conversion carry correlations that power quantum cryptography, quantum teleportation, and quantum computing. The monogamy of entanglement guarantees that those correlations cannot be secretly shared with an eavesdropper, making security physics-guaranteed rather than computationally hard. The Hong-Ou-Mandel dip quantifies the photon indistinguishability required by these protocols. And the no-communication theorem ensures that apparent non-locality does not permit faster-than-light signaling, thereby maintaining compatibility with special relativity despite the paradoxes the phenomenon seems to entail.

The next chapter takes a step outward from the laboratory toward the cosmos, exploring how quantum light is being used to sense and image with precision no classical technique can match. From gravitational wave detectors to quantum-enhanced microscopes, the same quantum-optical principles developed throughout this book now underlie the most sensitive measurement instruments humanity has ever constructed. The journey from a single entangled photon pair to an observatory that detects the merger of two black holes a billion light-years away is shorter than it might appear, and the physics connecting them is the same physics you have just learned.

8 Quantum Sensing and Imaging: Seeing the Unseeable

Opening Scenario

The news alert arrives on a Tuesday morning while most people are still drinking coffee. Scientists at LIGO, the Laser Interferometer Gravitational-Wave Observatory, have announced another detection: two neutron stars collided roughly 130 million years ago, and the ripple that collision sent through the fabric of spacetime passed through Earth last night. The signal lasted less than two seconds. It was buried far below the noise floor of any classical instrument. And yet LIGO found it, measured it precisely, and pinpointed its origin in the sky to within a fraction of a degree. The key ingredient was not bigger mirrors or a quieter building. It was a beam of squeezed light, a form of light whose quantum noise had been deliberately redistributed so that the measurement the scientists cared about became almost eerily precise. Without that squeezed light, the neutron-star merger would have slipped past unnoticed.

Or consider a hospital corridor in a pediatric neurology unit. A small child wearing what appears to be a bicycle helmet fitted with sensor pods sits and watches a cartoon. There is no loud banging, no tight cylindrical bore, no requirement to lie perfectly still. The sensors in that helmet are quantum magnetometers that measure the magnetic fields generated by individual clusters of neurons firing in the child's brain. The fields are extraordinarily faint, roughly one hundred billion times weaker than a refrigerator magnet, yet the sensors resolve them with millisecond timing and millimeter-scale spatial detail. No cryogenic cooling is required; the sensors run at room temperature, driven by laser light and quantum physics.

Both scenes are real, not speculative. LIGO has run with squeezed-light injection since 2019. Wearable quantum brain scanners have already produced clinical-quality recordings in human subjects. What unites them is a single insight: the quantum nature of light is not an esoteric laboratory curiosity but a practical engineering resource that can push measurement precision beyond limits once thought absolute. This chapter is about where those limits come from, how quantum optics crosses them, and what that crossing makes possible.

Why It Matters

Measurement is the foundation of all science and most engineering. Every clinical diagnosis, every navigation fix, every structural safety inspection, every astronomical discovery begins with a measurement trusted to be accurate. For most of history, the limit on precision was technological: better instruments, more carefully built, in quieter environments, produced better numbers. By the late twentieth century, the most demanding instruments began running into a different kind of limit, not a technological one but a physical one rooted in the quantum nature of light. That limit, called the standard quantum limit or shot-noise limit, is the measurement precision ceiling set by the inherent randomness in how individual photons arrive at a detector. Improving beyond it requires not better mechanics but a rethinking of the quantum state of the light itself.

Quantum sensing is the field built on exactly that rethinking. Its tools include squeezed light (which redistributes quantum noise away from the measurement axis), entangled photon pairs (which enable images formed from photons that never touched the object), laser-cooled atomic clocks (which tick at optical frequencies hundreds of thousands of times higher than their microwave predecessors), and spin-sensitive diamond defects (which map magnetic fields at nanometer scales and room temperature). The practical stakes are already visible: gravitational-wave astronomy listening to the universe at new depths, brain scanners freed from cryogenics, navigation systems that could operate without GPS, and range-finding sensors that detect targets invisible to classical instruments.

The Shot-Noise Floor and Why Classical Sensors Hit It

Every optical measurement ultimately depends on counting photons. A light-based sensor, whether a microscope, an interferometer, or a range finder, works by directing photons toward a target and observing the reflected signal. The precision of the result depends on how accurately the counts of returning photons are interpreted. And here is the unavoidable complication: photons do not arrive at a detector in a perfectly uniform stream. Even from the most stable laser, photons arrive randomly, following what statisticians call Poisson statistics. Think of raindrops falling at a steady average rate: the average is steady, but the individual drops cluster and spread irregularly. If you were trying to measure something subtle by timing the drops, that irregularity would limit how finely you could resolve it.

This irregularity in photon arrivals generates what physicists call shot noise, named by analogy to the random rattle of buckshot. The shot noise in a measurement decreases as the square root of the number of photons used; doubling the precision in this way requires quadrupling the photon count. In many situations, this is perfectly acceptable; you use more light. But in several important applications, increasing light intensity is not an option. Fluorescence microscopy imaging a living cell with more photons damages the proteins it is trying to observe. A gravitational-wave detector cannot raise laser power without distorting its mirror surfaces. A night-vision sensor cannot flood a target with extra photons without revealing itself. The standard quantum limit, the precision ceiling imposed by shot noise alone, is therefore not a minor annoyance in these contexts but a hard frontier.

Diagram 8.1 - The Shot-Noise Limit and Poisson Statistics of Photon Arrival

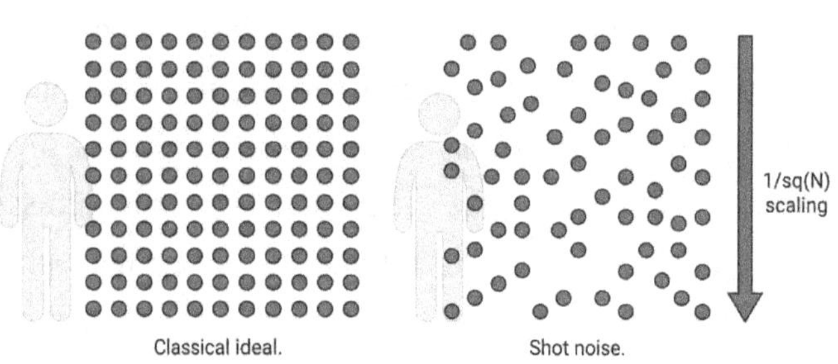

The standard quantum limit, often abbreviated SQL, can only be crossed by changing the quantum state of the light itself. The noise in a measurement is not an accidental feature of imperfect equipment; it is an intrinsic property of the quantum state being used. Different quantum states carry different noise distributions. Some states concentrate noise in dimensions you do not care about and reduce it in the dimension you do care about. Those states are the entry point to sensing beyond the shot-noise floor, and engineering them is the central task of quantum-enhanced metrology, the science of making measurements with precision beyond what classical physics permits.

Understanding the SQL also clarifies why most practical measurement improvements over the last century were achieved simply by better engineering, more stable mounts, quieter electronics, and thermally isolated environments. Those improvements attacked technical noise sources that sat well above the SQL. Once those technical noise sources are reduced below the SQL, continued improvement requires crossing into quantum-optical territory: squeezed states, entangled probes, and quantum-correlations-based detection schemes. Many of the most demanding instruments in science and technology have now reached precisely this point, which is why quantum sensing has transitioned from a curiosity to a necessity.

Quantum-Enhanced Sensing and the Heisenberg Limit

If the standard quantum limit is one fundamental floor, there is a deeper floor underneath it called the Heisenberg limit. When you use N photons in a classical measurement, each carries independent noise, and those noises average together, producing the familiar one-over-square-root-of-

N improvement described above. The Heisenberg limit represents the absolute best precision that quantum mechanics allows for any measurement made with exactly N photons, however cleverly the light is engineered.

The Heisenberg limit scales as one divided by N rather than one divided by the square root of N. For large N, this difference is dramatic: one thousand photons at the shot-noise limit yield about thirty-two times the precision of a single photon; one thousand photons at the Heisenberg limit could, in principle, yield a thousand-fold improvement. Reaching the Heisenberg limit requires creating highly entangled states called NOON states, a name that captures the quantum superposition involved: N photons simultaneously in a state of all-being-in-one-path AND all-being-in-the-other-path, existing in both possibilities at once until measured. NOON states are extraordinarily fragile; any stray interaction with the environment destroys the entanglement and eliminates the advantage.

Diagram 8.2 - Shot-Noise Limit vs. Heisenberg Limit: Precision Scaling Comparison

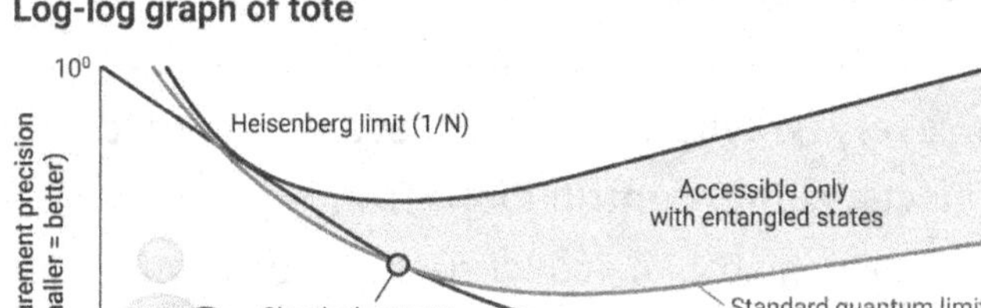

Entangled photons let sensors close the gap toward the ultimate quantum precision limit.

Between the standard quantum limit and the Heisenberg limit lies an intermediate territory where squeezed light operates. Squeezed light, introduced in Chapter 6, is a quantum state of light in which uncertainty has been redistributed between two complementary properties: amplitude and phase. A phase-squeezed beam has less phase noise than a classical beam of the same average intensity, at the cost of more amplitude noise. By injecting phase-squeezed light into a phase-sensitive interferometer, you reduce the measurement noise without violating the uncertainty principle; the extra noise lands in amplitude, which the interferometer does not measure. This is a trade-off the physics permits, and exploiting it is what squeezed-light sensing does.

Squeezed-light enhancement is deployed, not theoretical. LIGO began injecting squeezed light into its detector arms in 2019, achieving a sensitivity improvement roughly equivalent to doubling the laser power without the thermal and mechanical complications that increased power would bring. Future LIGO generations plan deeper squeezing as a primary route to increased reach into the cosmos. Every neutron-star merger, every black-hole collision, every gravitational-wave event recorded since 2019 reflects some contribution from quantum noise engineering of light: a concrete demonstration that quantum sensing has moved from the physics laboratory to the forefront of observational astronomy.

Diagram 8.3 - LIGO Squeezed-Light Interferometer Layout

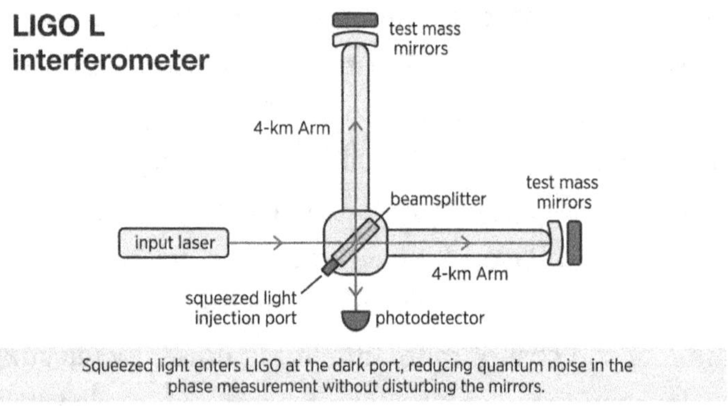

Squeezed light enters LIGO at the dark port, reducing quantum noise in the phase measurement without disturbing the mirrors.

Ghost Imaging: Forming Images with Photons That Never Touched the Object

One of the strangest demonstrations in all of quantum optics involves forming an image of an object using photons that never interacted with it. The technique is called ghost imaging, and it exploits entanglement in a way that has no classical analog. The starting point is an entangled photon pair source, typically a nonlinear crystal where a single pump photon enters and exits as two lower-energy photons whose directions, frequencies, and polarizations are correlated. These two photons are called the signal and idler photons.

In a ghost imaging setup, the signal photon is directed toward the object. It passes through or is reflected by the object, carrying information about its spatial structure. But the detector used to detect the signal photon has no spatial resolution; it is a simple bucket detector that only registers whether a photon arrived. The idler photon goes in a completely different direction, one that never brings it near the object. A high-resolution spatial camera detects the idler photon. By correlating the bucket detector events with the camera positions, a spatially resolved image of the object is gradually assembled entirely from idler photons that never touched it.

Diagram 8.4 - Ghost Imaging Setup with Entangled Photon Pairs

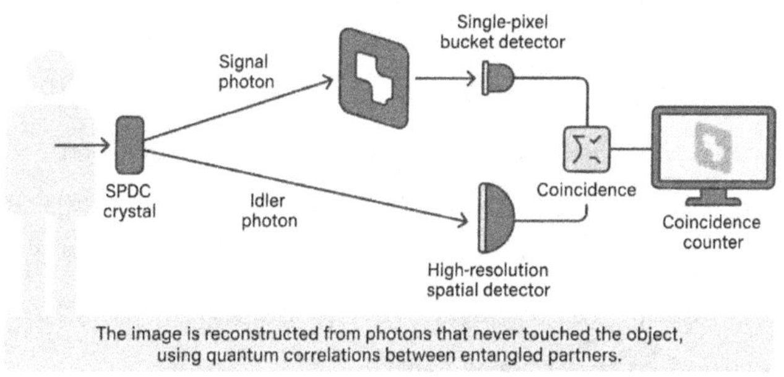

The practical utility of ghost imaging lies in wavelength flexibility. Because the photons that touch the object do not need to be the ones forming the image, you can illuminate the object with infrared or terahertz light that penetrates materials opaque to visible wavelengths, while forming the actual image using visible-wavelength idler photons on an inexpensive, high-resolution silicon camera. High-resolution infrared cameras are expensive and technically challenging to manufacture; high-resolution visible-light cameras are consumer-grade technology. Ghost imaging, therefore, enables infrared and terahertz imaging at the resolution and cost of visible-light systems, with applications in imaging through fog, tissue, and turbid media.

A closely related technique called quantum illumination applies the same logic to detection rather than imaging. An entangled photon pair is generated: one photon is sent toward a potential target in a noisy environment, while its partner is retained locally. The returning photon (if any) is compared against the retained partner using quantum-correlation-based processing. Even when the environment destroys most of the entanglement, residual quantum correlations provide a signal-to-noise advantage over classical detection. Quantum illumination is being explored seriously for low-observable target detection and for medical imaging contexts where the background is bright, and the signal of interest is faint.

NOON States and the Path to Super-Resolution Imaging

Ghost imaging demonstrates how entanglement can form images in novel ways. A different class of quantum optical experiments asks whether entanglement can improve the fundamental resolution limit of an optical microscope or telescope. The answer, in principle, is yes, and the quantum states responsible are the NOON states introduced in the discussion of the Heisenberg limit. To understand why they help, it helps to be clear about what limits classical resolution in the first place.

Any imaging system uses a lens or mirror to collect light and form an image. Wave optics imposes a ceiling on the finest detail that can be resolved: Two point sources closer than a specific distance (the Rayleigh resolution limit, which depends on the wavelength of light and the size of the lens) cannot be distinguished because their diffraction patterns overlap. This is the diffraction limit, and it has constrained optical microscopy for more than a century. NOON states that it circumvents it through a different mechanism. A NOON state of N photons does not produce a single fringe at the standard wavelength spacing; it produces N fringes in the same space. This N-fold compression of the fringe pattern corresponds to an N-fold improvement in spatial resolution: features that classical optics blurs together become distinguishable in the NOON state fringe pattern.

Diagram 8.5 - NOON State Interferometer and Sub-Diffraction Fringe Pattern

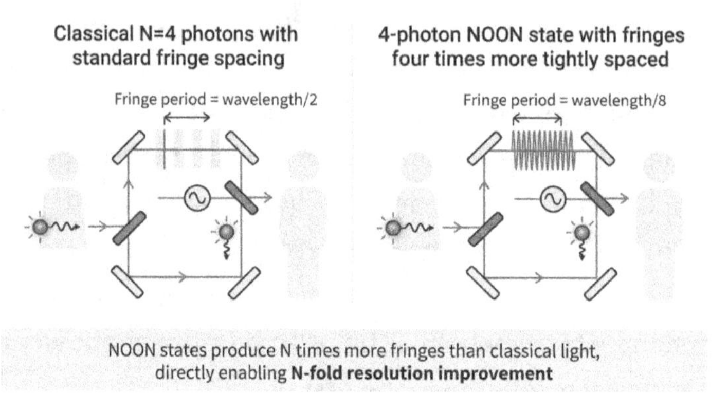

The practical challenges are formidable. Producing NOON states with N beyond about five or six photons requires generating and controlling multi-photon entanglement with extraordinary reliability, because any lost photon destroys the quantum correlations and eliminates the advantage. Laboratory demonstrations have confirmed the principle with two-photon and four-photon NOON states in controlled conditions. Whether NOON-state microscopy becomes a routine tool for biologists and materials scientists depends on advances in generating NOON states in practical, lossy environments, which is an active area of research. The physics is unambiguous; the engineering challenge remains open. What makes this area worth watching is that the biological sciences are among the most photon-budget-constrained disciplines in all of measurement science: the specimens are fragile, the structures of interest are tiny, and the signals are faint.

Any technology that extracts more information from fewer photons has a natural home in biophysics.

A more accessible near-term route uses the second-order statistical correlations of entangled photon pairs rather than high-order NOON states. Techniques sometimes grouped under the name quantum-inspired super-resolution exploit the photon-pair coincidence statistics to distinguish between point sources separated below the classical Rayleigh limit. These approaches tolerate loss better and require fewer exotic quantum states, making them realistic candidates for near-term deployment in fluorescence microscopy. The field is young, and new results in quantum-enhanced imaging appear regularly as the quantum optics and microscopy communities increasingly collaborate.

Optical Atomic Clocks: The Most Precise Timekeepers Ever Built

The atomic clocks aboard GPS satellites are what make sub-meter navigation possible. Without their timing precision, the triangulation of position from satellite signals would be too imprecise to be useful. But GPS-grade atomic clocks are already a generation behind the current state of the art. The frontier of clock technology has moved from the microwave frequencies of cesium atomic clocks into the optical frequency regime, producing instruments of such extraordinary accuracy that they are forcing a rethinking of how time itself is defined.

An atomic clock uses the precise, quantum-mechanically fixed frequency of a transition between two energy levels in an atom as its timekeeping reference. In the cesium atomic clocks that have defined the SI second since 1967, the relevant transition is a microwave transition at exactly 9,192,631,770 cycles per second. Every cesium atom in the universe undergoes this transition at the same frequency, making it an ideal and reproducible reference. The precision of any clock improves with the frequency of its reference: a higher frequency means more ticks per second, and more ticks per second means any error in reading a tick translates to a smaller error in time. Optical atomic clocks take this principle to an extreme by using transitions in the visible or near-infrared regime, at frequencies hundreds of thousands of times higher than those of microwave transitions.

Diagram 8.6 - Optical Lattice Clock with Laser-Cooled Atoms

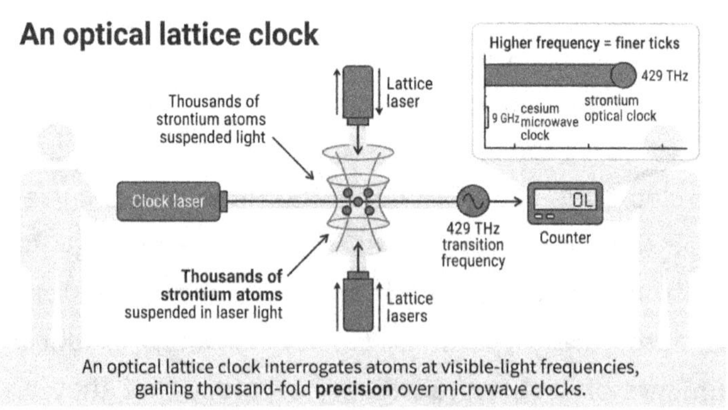

An optical lattice clock interrogates atoms at visible-light frequencies, gaining thousand-fold **precision** over microwave clocks.

Strontium-87 has an optical transition at approximately 429 terahertz, meaning the clock ticks more than four hundred trillion times per second. The best optical lattice clocks now demonstrated are accurate to better than one second over thirty billion years, exceeding the age of the universe by more than a factor of two. Achieving this requires laser cooling of the atoms to within a millionth of a degree of absolute zero, trapping them in an optical lattice (a standing-wave pattern of laser light that forms an array of microscopic wells holding individual atoms), and interrogating the transition with a clock laser of extraordinary spectral purity. Every stage of this process is a quantum optical engineering feat.

The applications of optical clocks extend well beyond timekeeping. Because gravity slows time (a well-confirmed consequence of general relativity), an optical clock at the bottom of a mountain ticks measurably more slowly than the same clock at the summit, even over elevation differences of just a few centimeters. This gravitational redshift is accessible with transportable optical clocks of sufficient accuracy, opening a new approach to geodesy, the science of measuring Earth's shape and gravitational field. A network of optical clocks could serve as a gravitational potential map of Earth with centimeter-level precision, with applications in earthquake monitoring, hydrological science, and the detection of underground geological structures of economic or security interest.

Quantum LiDAR and Standoff Detection

LiDAR, standing for Light Detection and Ranging, is the technology behind the three-dimensional mapping sensors on self-driving cars, orbiting terrain-mapping satellites, and robotic navigation systems. The basic principle is simple: fire a laser pulse toward a target, time how long the reflected pulse takes to return, and convert that round-trip time to a distance. Classical LiDAR is impressively capable, but it faces the same fundamental limits as all optical measurements: at low photon fluxes, shot noise limits ranging precision; in cluttered environments, the signal-to-noise ratio for faint targets drops below the detection threshold; and with coherent pulses, certain targets can take countermeasures against detection.

Quantum LiDAR, sometimes called quantum lidar, addresses these limitations through several strategies. The simplest is single-photon LiDAR, which uses detectors sensitive enough to register individual photons, called single-photon avalanche diodes or SPADs, rather than requiring a bright reflected pulse. Single-photon LiDAR illuminates entire scenes with a handful of photons per pixel per frame, recovering three-dimensional structure from the coincidence statistics of signal photons against a known timing reference. The challenge is separating signal photons from random background photons, and quantum-optical ideas about temporal and spectral correlations help: a signal photon from a known pulse can be gated in both time and wavelength against background radiation that lacks those correlations.

Diagram 8.7 - Quantum LiDAR vs. Classical LiDAR: Signal Recovery at Low Photon Flux

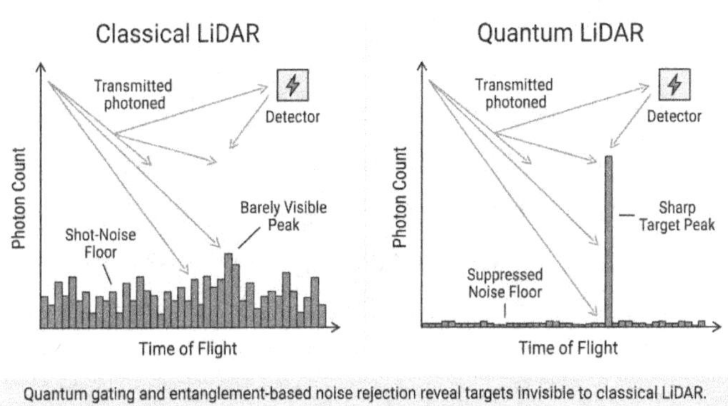

Quantum gating and entanglement-based noise rejection reveal targets invisible to classical LiDAR.

More sophisticated quantum lidar schemes use entangled photon pairs in a configuration analogous to quantum illumination. One photon of each pair is sent toward a potential target; its entangled partner is retained at the transmitter. If a photon returns from the target, it is compared with the retained idler via a joint quantum measurement that tests for entanglement-derived correlations. Even after the round-trip through the environment destroys most of the entanglement, the residual correlations survive at a level sufficient to distinguish a true target return from a background photon arriving at the same time. The signal-to-noise improvement is largest exactly where classical lidar struggles most: dim targets in bright, noisy backgrounds.

Quantum-enhanced ranging also benefits from the ultra-narrow coincidence window of entangled photon pairs. Because both photons of a pair are generated simultaneously within a time window set by the pump laser coherence, they provide a timing reference much more precise than the pulse width of a classical laser. Any photon not correlated with the retained idler within that narrow window is rejected as noise. Defense research agencies in multiple countries are actively funding quantum lidar programs, and integrated silicon photonic platforms are already demonstrating on-chip generation of entangled pairs at rates compatible with practical ranging applications.

OPM-MEG and Quantum Sensing in Medicine

The human brain is an electrical machine. Hundreds of billions of neurons fire in coordinated patterns, sending currents down axons and across synapses. Those currents generate magnetic fields. The fields are extraordinarily faint, typically tens to hundreds of femtoteslas (a femtotesla is one millionth of one billionth of a tesla; the Earth's field is about a quadrillion times stronger). Measuring those fields in real time would give clinicians and neuroscientists a direct window into brain activity with millisecond timing. The challenge is building a sensor sensitive enough to detect faint signals while filtering out the surrounding environment.

For decades, the only practical brain magnetic imaging technology used superconducting quantum interference devices, called SQUIDs. A SQUID is a loop of superconducting wire interrupted by two thin insulating barriers, called Josephson junctions, whose quantum interference pattern shifts detectably in the presence of a magnetic field. SQUIDs are sensitive instruments, but they have a critical limitation: superconductivity requires temperatures near absolute zero, which means immersion in liquid helium inside a heavy cryogenic dewar. The patient must hold completely still with their head inside a rigid, helmet-like cryogenic apparatus. This excludes infants, young children, patients with movement disorders, and anyone whose clinical condition prevents sustained stillness from being studied with any practical convenience.

Diagram 8.8 - Optically Pumped Magnetometer Principle

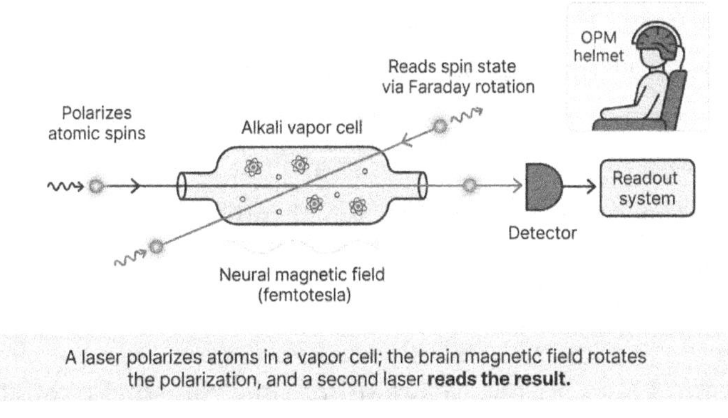

A laser polarizes atoms in a vapor cell; the brain magnetic field rotates the polarization, and a second laser **reads the result.**

Optically pumped magnetometers, abbreviated OPMs, offer brain magnetic imaging without cryogenics. An OPM uses laser light to prepare atoms in a specific quantum spin state, a process called optical pumping, and then monitors how that spin state evolves in the presence of the ambient magnetic field. The atoms are placed in a coherent quantum superposition of spin states, and the rate at which this quantum coherence precesses directly encodes the field strength. A second laser beam reads out the spin state via the Faraday effect, the rotation of the beam's polarization angle by the magnetized atomic vapor, or by monitoring the fluorescence intensity of the atoms. Modern OPMs have reached sensitivities comparable to or exceeding those of SQUID magnetometers while operating at room temperature.

When arranged in a wearable helmet array resting directly on the scalp, OPM-based magnetoencephalography (OPM-MEG) offers both better sensitivity and better spatial resolution than cryogenic SQUID systems. The reason is proximity: the closer a magnetic sensor sits to the source of the field, the stronger the signal and the finer the spatial detail that can be distinguished. Pressing an OPM sensor against the scalp surface is the magnetic sensing equivalent of pressing your ear against a wall to hear a sound; the cryogenic helmet, which must stand off from the head by the thickness of the insulation, cannot achieve the same intimacy.

Diagram 8.9 - OPM-MEG Brain Imaging: Sensor Proximity Advantage

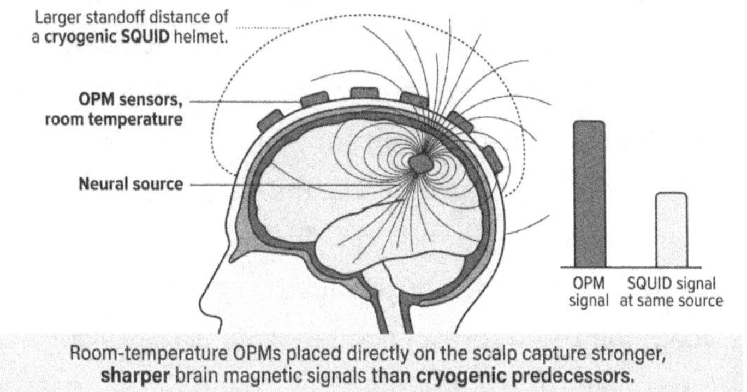

Room-temperature OPMs placed directly on the scalp capture stronger, **sharper** brain magnetic signals than **cryogenic predecessors.**

Clinical results with OPM-MEG are already striking. Epileptic seizure foci have been localized with greater precision than prior SQUID systems, improving surgical planning for patients who need resective surgery. Resting-state brain networks have been mapped in children as young as two years, an age group that SQUID-MEG essentially could not image because of the requirement for sustained stillness. Neurodegenerative signatures of Parkinson's and Alzheimer's disease produce characteristic changes in the brain's oscillatory magnetic signals that OPM-MEG is beginning to resolve with enough consistency to consider as diagnostic biomarkers. Researchers studying language, memory, and social cognition have begun using OPM-MEG to record brain activity during genuinely naturalistic tasks: conversation, walking, and reaching for objects. The window onto human brain function that this technology opens is qualitatively wider than anything that preceded it.

The NV center magnetometer, built around a point defect in diamond where a nitrogen atom sits adjacent to a missing carbon atom (the vacancy), represents a complementary quantum sensing approach at even finer spatial scales. This nitrogen-vacancy defect can be initialized into a specific quantum spin state using green laser light and then read out via the intensity of its red fluorescence, which depends on the spin state. Because the spin state is exquisitely sensitive to nearby magnetic fields, mapping the fluorescence intensity across a surface yields a magnetic field map at nanometer spatial resolution. NV center magnetometers are already used in materials science to image magnetic domain walls in thin films, map current flows in superconductors, and detect the magnetic fields of individual living bacteria. They represent one of the most versatile room-temperature quantum sensors yet developed.

Diagram 8.10 - NV Center in Diamond: Optical Spin Readout

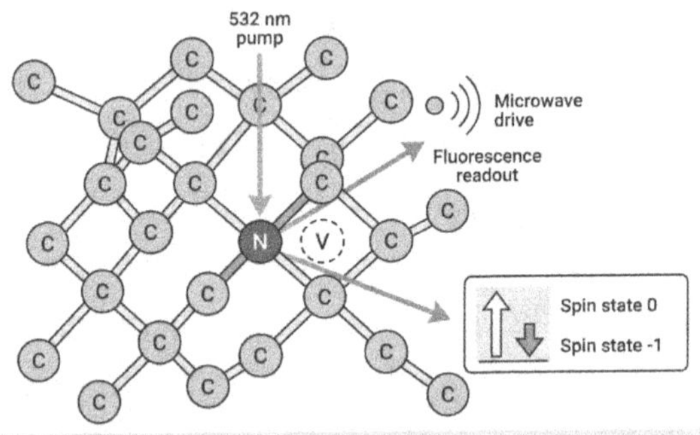

A nitrogen-vacancy defect emits light whose brightness encodes its quantum spin state, creating a nanoscale magnetic sensor.

What This Means For You

For engineers and practitioners in photonics, imaging, or instrumentation, the technologies surveyed in this chapter define a frontier moving rapidly from demonstration to deployment. The squeezed-light infrastructure now operating inside LIGO is a proof of concept for large-scale quantum-enhanced interferometry, and the engineering lessons learned there are being absorbed by teams building next-generation optical clocks, fiber-sensing systems, and quantum-enhanced spectrometers. Quantum noise engineering is becoming a practical design option for high-precision measurement instruments of any kind, not a theoretical aspiration.

For those in medical technology, OPM-MEG is perhaps the most immediately consequential development. Systems are already entering early clinical use, and the pipeline of improved sensor designs, better noise-rejection algorithms, and miniaturized form factors is moving quickly. The clinical case is strong: millisecond time resolution, millimeter spatial resolution, room-temperature operation, and the ability to image naturally moving patients address genuine limitations of both fMRI (which offers spatial resolution but poor time resolution) and SQUID-MEG (which offers time resolution but requires cryogenics and strict stillness). Healthcare organizations evaluating neuroimaging capital should be actively tracking this field.

For those in defense, aerospace, and navigation, two developments converge. Optical atomic clocks being miniaturized for portable deployment in submarines, aircraft, and ground vehicles would provide GPS-independent timing and positioning at performance levels exceeding GPS by orders of magnitude, a decisive advantage in environments where satellite signals are unavailable or have been jammed. Quantum lidar and quantum illumination offer target detection and ranging in cluttered or adversarial environments where classical active sensors are vulnerable to countermeasures. Both areas are receiving substantial government funding globally, and the transition from laboratory demonstrations to engineered systems is already underway.

For the technically curious reader without a direct professional stake in any of these fields, the broader lesson of quantum sensing is worth holding onto. Improving a measurement has historically meant building a better instrument: more precise, more stable, better shielded. Quantum sensing demonstrates that the next generation of improvements requires something different: the deliberate engineering of the probe's quantum state. That shift in thinking, from 'build a better instrument' to 'engineer a better quantum state,' is the same shift that runs through every chapter of this book and through every branch of quantum technology. It is the conceptual move that separates the quantum era from everything that came before it.

Diagram 8.11 - Sub-Shot-Noise Interferometry with Squeezed Input

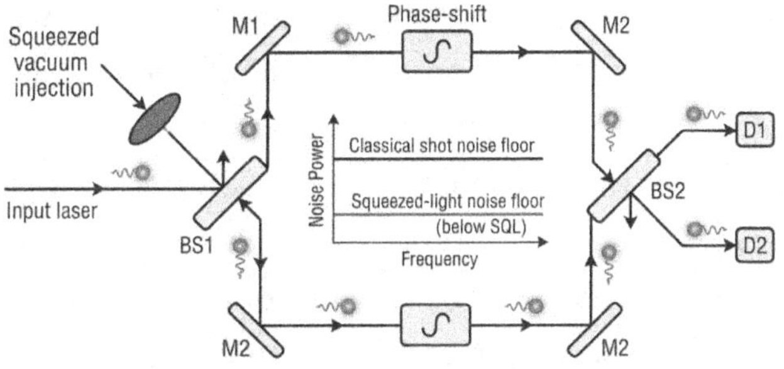

Injecting squeezed light suppresses quantum noise in the interferometer below the standard quantum limit.

Takeaway

The quantum sensing revolution is built on a single counterintuitive truth: the noise that limits the precision of optical measurements is not a fixed property of the instruments but rather a property of the light's quantum state. Change the state, and you change the noise. This insight, which became rigorous theory in the early 1980s and engineering practice in the 2010s, has already produced instruments of unprecedented capability: gravitational-wave detectors listening to mergers a hundred million light-years away, atomic clocks more accurate than the age of the universe, brain scanners freed from cryogenics, and lidar systems detecting targets invisible to classical sensors.

The standard quantum limit, the precision ceiling set by Poisson photon statistics in any classical measurement, is not the last word. Squeezed light crosses the SQL by redistributing quantum uncertainty from the measured dimension into an unmeasured one, without violating any physical law. Entangled photon pairs enable ghost imaging and quantum illumination by exploiting correlations that classical light cannot reproduce. NOON states point toward the Heisenberg limit, where precision scales with photon number rather than its square root. Optical atomic clocks, by interrogating atomic transitions at visible-light frequencies, achieve timing precision beyond the lifetime of the universe. And OPM and NV center magnetometers bring room-temperature sensitivity to magnetic field strengths so faint that classical instruments cannot approach them outside a cryogenic environment.

What ties all of these together is a consistent position: the quantum nature of light is not a problem to be engineered around but a resource to be deliberately harnessed. Squeezed light is a resource shaped for interferometry. Entangled pairs are a resource shaped for imaging and detection. Laser-cooled atoms are a resource shaped for timekeeping. NV centers are a resource shaped for nanoscale magnetometry. The field is young enough that its most important applications may not yet have been conceived, but mature enough that several of its instruments are already rewriting what science and medicine can observe. The unseeable is becoming seeable, one quantum state at a time.

Diagram 8.12 - Gravitational-Wave Detection Event: Squeezed vs. Classical Noise Floor

Executive technical infographic-style, time series signal plot

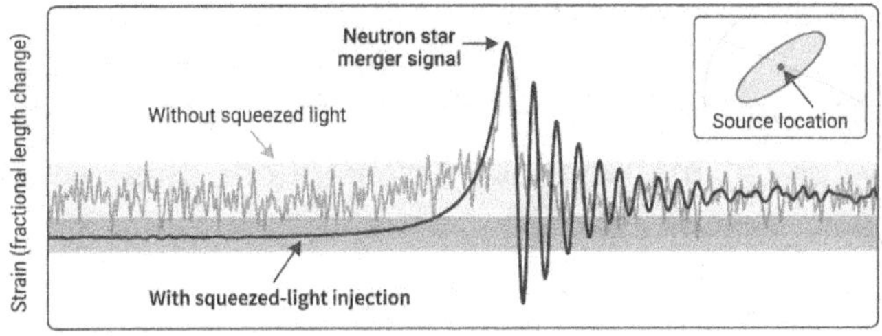

Squeezed light lifts the neutron-star merger signal above the noise floor, enabling detection that classical sensitivity could not achieve.

9 Quantum Communication and Cryptography: Unbreakable by Physics

Opening Scenario

You tap the banking app on your phone, type your password, and watch the small padlock icon snap closed in the corner of the screen. For most people, that padlock is visual wallpaper, a reassuring detail that fades into familiarity after the first few uses. But ask what it actually guarantees. That padlock indicates that the connection between your phone and your bank has been encrypted with a mathematical key that only your device and the bank's server hold. Nobody sitting on the same Wi-Fi network can read your balance or intercept your transfer. The system works and has worked reliably for decades. It is the backbone of online commerce, digital banking, government communications, and private messaging for billions of people worldwide.

The quiet problem is that the padlock's security rests entirely on a mathematical assumption: that certain arithmetic problems are so hard that no computer, even in a practical timeframe, could crack them. The most widely used public-key encryption systems rely on the fact that multiplying two large prime numbers is trivial, but working backward to recover those primes from their product is computationally expensive. That assumption has held for decades. But quantum computers, which exploit the strange behaviors of quantum mechanics, can solve exactly this problem efficiently using an algorithm called Shor's algorithm. A sufficiently large fault-tolerant quantum computer would reduce today's padlock to an open latch in a matter of hours.

This chapter is about what comes after that padlock. Quantum key distribution, often abbreviated QKD, replaces mathematical hardness with physical law as the foundation of cryptographic security. Instead of encoding a secret key in a mathematical puzzle, QKD encodes it in the quantum states of individual photons, particles of light. The security guarantee does not rest on assumptions about computational difficulty. It rests on the structure of the universe: the rules of quantum mechanics make it physically impossible for any eavesdropper to intercept those photons without leaving a detectable trace. It does not matter how powerful the attacker is or what algorithms they develop in the future. The laws of physics do not have software updates.

Why It Matters

The race to deploy quantum-safe cryptography is not about protecting tomorrow's secrets. It is about protecting today's secrets from tomorrow's computers. Security professionals refer to a threat as "harvest now, decrypt later," sometimes abbreviated as the "harvest-now-decrypt-later" attack. In this scenario, an adversary intercepts and stores today's encrypted traffic, even though they cannot read it yet. A decade or two from now, when a powerful quantum computer exists, they run Shor's algorithm against the archived data and decrypt it retroactively. Any organization whose data needs to remain secret for more than ten years faces real and present risk.

Quantum key distribution offers a physics-based defense against this threat. Running alongside it is a complementary field called post-quantum cryptography, abbreviated PQC, which involves developing mathematical algorithms that resist attacks by even quantum computers. Both approaches matter, and they work together rather than competing. QKD holds a unique position because its guarantees are unconditional in a physical sense, not contingent on any mathematical conjecture remaining unbroken. Understanding QKD also opens a window onto a larger vision: the quantum internet, a planetary network in which quantum entanglement can be shared between any two points on Earth.

Why Classical Encryption Is Vulnerable and What Quantum Communication Changes

To appreciate what quantum communication offers, it helps to understand exactly where classical encryption is vulnerable. The dominant system in use today is called RSA, named for its inventors Rivest, Shamir, and Adleman. RSA is a public-key cryptosystem, which means it uses two mathematically related keys: a public key that anyone can use to encrypt a message, and a private key that only the recipient holds for decryption. The relationship between them depends on the difficulty of factoring the product of two very large prime numbers. Primes are numbers divisible only by themselves and one. Classical computers cannot factor a 2048-bit RSA key in any reasonable span of time. That difficulty with factoring is the lock on the door.

Shor's algorithm, published by mathematician Peter Shor in 1994, changes this entirely. Running on a quantum computer, Shor's algorithm factors large numbers in polynomial time, meaning the computation time grows only modestly as the number grows larger, rather than exponentially. A quantum computer running Shor's algorithm could crack a 2048-bit RSA key in hours. The same threat applies to elliptic-curve cryptography, which powers much of mobile and web security. The mathematical hardness assumptions underlying both systems collapse under quantum attack.

Quantum key distribution abandons the mathematical approach entirely. Its security derives from the laws of quantum physics, specifically from two properties: the no-cloning theorem, which prohibits perfectly copying an unknown quantum state, and the measurement postulate, which guarantees that measuring a quantum system disturbs it. These are not engineering constraints or computational limitations. They are verified properties of reality that no algorithm, classical or quantum, can circumvent.

Diagram 9.1 - Classical vs. Quantum-Secure Padlock

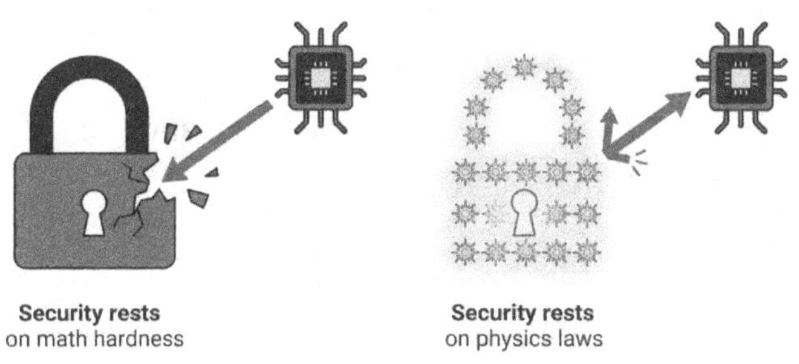

Classical encryption relies on hard math; quantum key distribution relies on the laws of physics.

The No-Cloning Theorem: Why Quantum States Cannot Be Copied

The security of quantum key distribution rests on a bedrock principle called the no-cloning theorem. In the classical world, copying is trivial: a bit can be read and duplicated without error or disturbance. An eavesdropper can silently and perfectly copy a classical signal, forwarding the original undisturbed while keeping a copy for analysis. This is precisely why classical cryptography cannot use physical transmission as its security foundation.

Quantum mechanics operates under fundamentally different rules. A photon used as a qubit, a quantum bit of information, carries information in its polarization, the orientation of its electric field oscillation. A photon can be polarized horizontally, vertically, at 45 degrees, or at 135 degrees, or it can exist in a superposition of these states simultaneously. The no-cloning theorem states that no physical operation can take an unknown quantum state and produce a perfect copy while leaving the original intact. This is not a statement about technological limitations. It is a mathematical consequence of the linear, reversible nature of quantum evolution, and no future hardware improvement can change it.

For cryptography, the no-cloning theorem is a gift from nature. Any eavesdropper, conventionally called Eve in cryptographic convention, who tries to intercept a photon carrying part of a quantum key cannot copy it and forward the original undisturbed. Any measurement Eve makes on the photon's quantum state perturbs it. That disturbance is detectable by Alice, the sender, and Bob, the receiver, when they compare a portion of their key bits over an ordinary classical channel. If Eve were active, her measurements would have introduced statistical errors above the expected noise level, and Alice and Bob would know the channel was compromised before using the key.

Diagram 9.2 - The No-Cloning Theorem and Eve

Quantum state Corruption

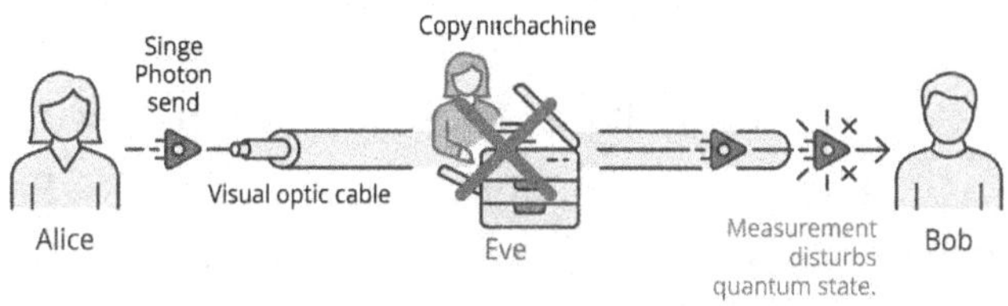

Any attempt by Eve to copy a quantum state inevitably damages it, revealing her presence.

BB84: The First and Most Important QKD Protocol, Step by Step

The most famous and widely deployed quantum key distribution protocol is BB84, named for its inventors Charles Bennett and Gilles Brassard and the year they published it: 1984. BB84 translates the abstract security guarantee of the no-cloning theorem into a practical step-by-step procedure for generating a shared secret key using photons.

Step one: Alice prepares photons. Alice randomly assigns each bit a value of 0 or 1, and for each bit, she also randomly chooses one of two measurement bases for encoding. In the rectilinear basis, horizontal polarization represents zero, and vertical polarization represents one.

In the diagonal basis, 45-degree polarization represents zero, and 135-degree polarization represents one. She encodes each bit as the corresponding polarization and sends that photon to Bob through a quantum channel, typically an optical fiber.

Step two: Bob measures. Bob receives each photon and also randomly chooses a basis, independently of Alice. If Bob chooses the same basis as Alice, his measurement yields the correct result. If he chooses the wrong basis, his result is essentially random. Step three: Basis reconciliation. Alice and Bob communicate over a classical public channel, announcing which basis they used for every photon but not their measurement outcomes. They discard all bits where their bases differed and keep the matching-basis bits, roughly half the total. This surviving set is called the sifted key.

Step four: Error checking. Alice and Bob publicly compare a random sample of their sifted key bits. If Eve intercepted photons, her wrong-basis guesses would disturb the photon states, and when she re-sent them to Bob, his measurements in the correct basis would have a 25 percent chance of recording the wrong value. Enough interceptions create a detectable spike in the error rate. Step five: Privacy amplification distills a shorter but provably secure final key from the remaining bits, making any partial information Eve gathered useless. Either she gathers significant information and generates detectable errors, or she intercepts few photons and gains negligible information.

Diagram 9.3 - BB84 Protocol Step by Step

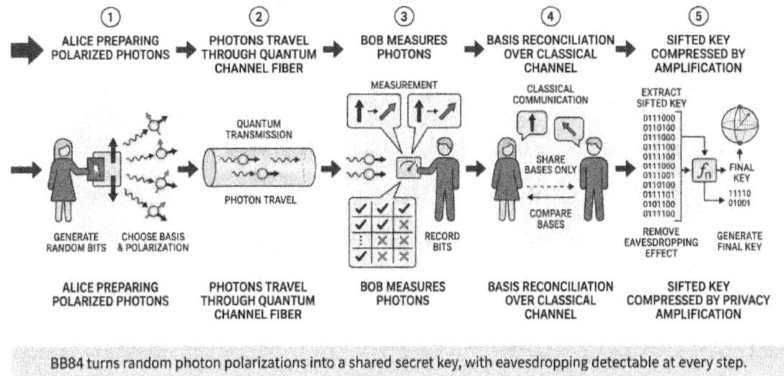

BB84 turns random photon polarizations into a shared secret key, with eavesdropping detectable at every step.

How Eavesdropping Is Detectable Through Quantum State Disturbance

The detection of eavesdropping in BB84 reveals something philosophically profound: the channel needs no locks, guards, or pre-shared secrets to protect the key in transit. The key protects itself because it is quantum mechanical. Any observation of a quantum system changes it, and that change is measurable.

Consider what Eve must do. She is somewhere along the fiber between Alice and Bob. She wants to learn the polarization of each photon, which encodes the key bit. But she does not know which basis Alice used for each photon. She must guess. If she guesses correctly, her measurement extracts the right value, and she resends an identical photon; Bob sees no error from that photon. If she guesses wrong, which happens about half the time on average, her measurement collapses the photon into a state in her basis rather than Alice's. The photon she forwards to Bob is prepared incorrectly, and when Bob measures it in the right basis, he has a 50 percent chance of getting the wrong answer.

The arithmetic is clean. Eve guesses the wrong basis half the time, and when she does, Bob gets the wrong bit half the time. That is one error for every four photons Eve intercepts, giving a 25 percent error rate. Real channels have some background noise, so the accepted threshold is typically around 11 percent. Any error rate above that triggers a key abort. Partial eavesdropping is also captured: fewer interceptions mean fewer errors but also less information gained, and privacy amplification accounts for the full trade-off between information gathered and errors introduced.

Diagram 9.4 - Eavesdropping Error Statistics in BB84

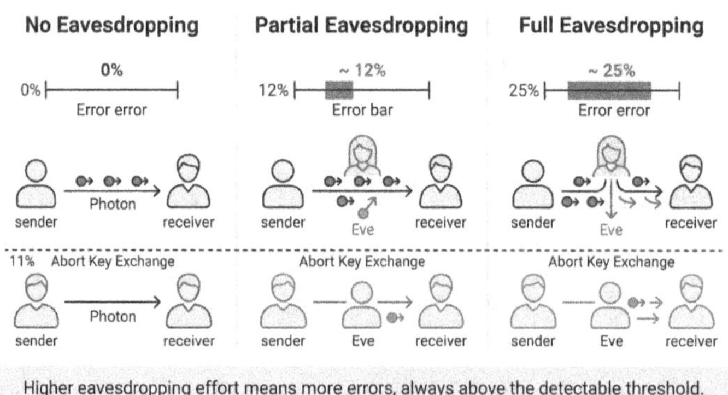

From Lab to Field: Decoy-State Protocol, MDI-QKD, and Twin-Field QKD

The theoretical security of BB84 is elegant, but real-world deployment is messier. Practical QKD systems face hardware imperfections that open security gaps not present in the ideal protocol. Two of the most consequential challenges are imperfect photon sources and detector vulnerabilities.

The ideal protocol requires a source producing exactly one photon per pulse. If two or more photons travel in the same pulse, an eavesdropper can intercept one extra photon while letting another reach Bob, gaining information without disturbing what Bob receives. This is called a photon-number-splitting attack. Most practical QKD systems use attenuated lasers rather than perfect single-photon sources. A Poisson distribution of photon numbers means some pulses will contain two or three photons, creating a real vulnerability. The decoy-state protocol solves this by having Alice transmit pulses at several different average photon intensities, called decoy states, interspersed randomly with signal pulses. By analyzing the intensity levels Bob successfully detects, Alice and Bob can estimate whether multi-photon pulses are being selectively exploited. The decoy-state protocol is now standard in commercial QKD hardware.

Detector vulnerabilities are a different attack surface. Strong classical light pulses can temporarily blind real single-photon detectors, and a sophisticated eavesdropper can exploit this to control what Bob thinks he detects. Measurement-device-independent QKD, known as MDI-QKD, was designed to eliminate this class of attacks. In MDI-QKD, both Alice and Bob send photons to an untrusted central measurement node. Security does not depend on that node being honest or on having uncompromised detectors. Because the measurements that could be attacked are performed by an assumed-adversarial party, there is nothing for a detector attack to exploit from Alice's or Bob's side.

Diagram 9.5 - MDI-QKD vs. Prepare-and-Measure QKD

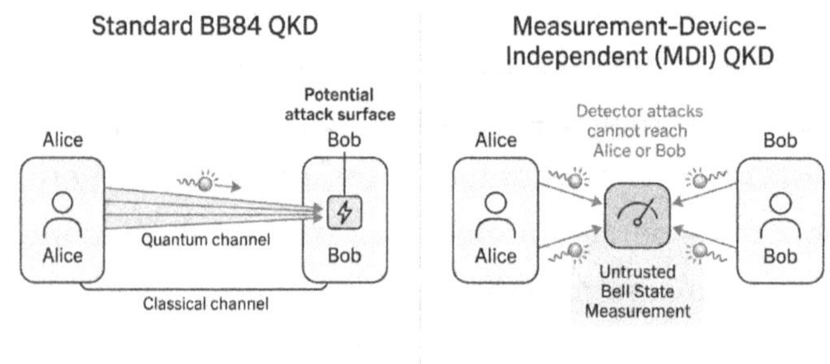

MDI-QKD removes the detector from the trusted zone, eliminating the most dangerous hardware attack surface.

Trusted Nodes, QKD Network Topology, and Metropolitan Quantum Networks

A single point-to-point QKD link is powerful. But real networks consist of many nodes connected by many links, and extending QKD across a network requires a way to relay keys across multiple hops. Quantum signals cannot be amplified, because amplification requires measuring and copying them, both of which are prohibited by the no-cloning theorem. The current practical answer is the trusted node architecture.

A trusted node, also called a trusted relay, is an intermediate node that shares QKD links with both its neighbors. When Alice wants to send a key to a distant Bob through a trusted node, Alice and the node run QKD to establish a shared key between themselves. The node and Bob then run QKD to establish a separate shared key. To relay Alice's key to Bob, the node applies a classical operation that combines the two keys and sends the result to Bob over a classical channel. Bob reverses the operation and recovers Alice's original key. The relay is classical, not quantum.

The security model of a trusted node network is weaker than a direct quantum link. The trusted node has access to all keys passing through it. If a trusted node is physically compromised, the adversary can read relayed keys. Real-world QKD networks using trusted nodes are therefore designed with careful physical security at each node: hardened facilities, restricted access, and tamper-evident hardware. China's national quantum communication backbone spans thousands of kilometers via a chain of trusted nodes and carries real keys for government and financial-sector applications. Tokyo, Geneva, and other cities have deployed similar networks. These are operational infrastructure, not prototypes.

Diagram 9.6 - QKD Network Topology with Trusted Nodes

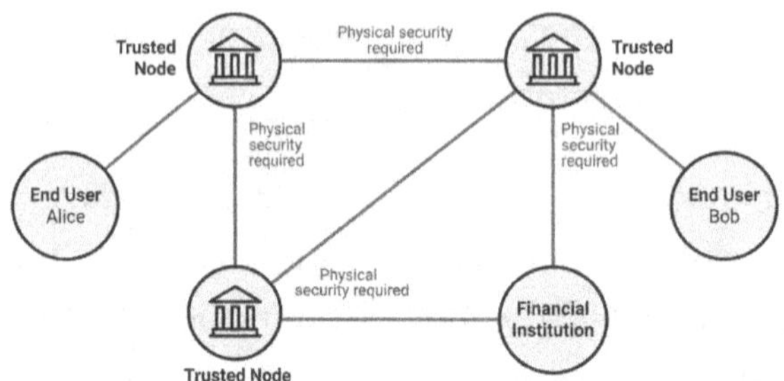

In today's QKD networks, trusted relay nodes extend reach at the cost of physical security at each intermediate site.

Satellite QKD and Intercontinental Quantum Communication

Fiber-based QKD has a fundamental distance ceiling. Photons are gradually absorbed and scattered in an optical fiber, and the probability of a photon arriving intact drops exponentially with distance. At typical telecom wavelengths in standard fiber, roughly half the photons are lost every 15 to 20 kilometers. At 200 kilometers, the loss is so severe that the key generation rate collapses to an impractical level. Without quantum repeaters, which are still under development, fiber QKD is effectively limited to city-scale or regional distances.

Satellite-based QKD circumvents this by using the sky as a channel. Photons traveling through the atmosphere and the vacuum of space experience far less loss than photons in fiber over equivalent distances. The effective absorbing column of the atmosphere is only about 10 kilometers for a vertically aimed beam, and the vacuum of space absorbs nothing. A satellite in low Earth orbit can relay quantum optical links between ground stations separated by thousands of kilometers without the photons ever traveling through that much fiber.

The landmark demonstration was China's Micius satellite, also known by its official designation QUESS (Quantum Experiments at Space Scale), launched in August 2016. In 2017, the Micius team demonstrated QKD between the satellite and two Chinese ground stations more than 1,200 kilometers apart. In 2018, they extended this to an intercontinental video call between scientists in China and Austria, with the encryption key distributed via satellite QKD. This was the first intercontinental quantum-secured communication link in history, spanning approximately 7,600 kilometers of ground separation.

Diagram 9.7 - Micius Satellite QKD

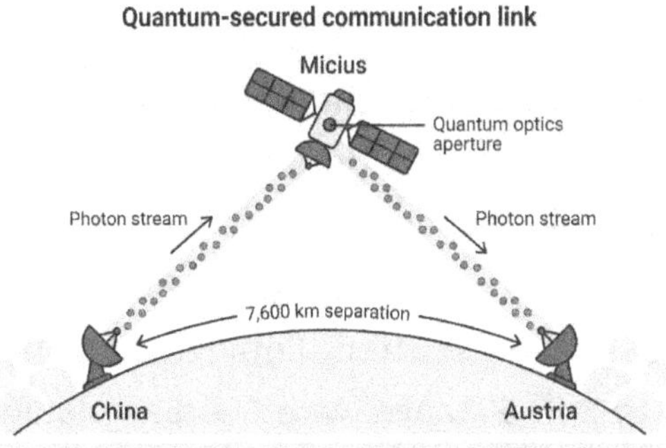

China's Micius satellite demonstrated the first intercontinental quantum-secured communication link in 2018.

The Quantum Repeater Problem and Its Solution Pathways

For a fiber-based global quantum network, the most critical missing piece is the quantum repeater. In classical networks, signal loss over long distances is solved by optical amplifiers, which boost a weakened signal by measuring and regenerating it. In a quantum network, this approach is forbidden. Measuring a quantum signal destroys its quantum properties, and copying it violates the no-cloning theorem. A fundamentally different approach is needed.

The quantum repeater uses two quantum techniques together: entanglement swapping and quantum memory. Entanglement swapping extends entanglement over longer distances by combining shorter entangled links. Imagine a chain of three nodes: Alice, a middle node called Charlie, and Bob. Charlie shares an entangled photon pair with Alice and a separate entangled pair with Bob. By performing a Bell state measurement on his two photons, one from each pair, Charlie causes Alice's photon and Bob's photon to become entangled, even though they have never directly interacted. Entanglement now spans the full Alice-to-Bob distance, with Charlie holding no copy of it.

Quantum memory is needed because entanglement swapping requires synchronization. Charlie needs Alice's photon and Bob's photon to be present at the same time. Without a way to hold a photon's quantum state while waiting for the other to arrive, the timing rarely aligns. A quantum memory accepts a photon, absorbs its quantum state into a material system such as cold atoms or a rare-earth-doped crystal, and releases it on demand. Key metrics are coherence time, the duration the memory holds a state before errors accumulate; retrieval efficiency, how reliably the stored state can be read back; and bandwidth, how many states the memory can hold simultaneously. Current devices achieve coherence times from milliseconds to seconds, with retrieval efficiencies of 50 to 80 percent.

Diagram 9.8 - Quantum Repeater Chain with Entanglement Swapping

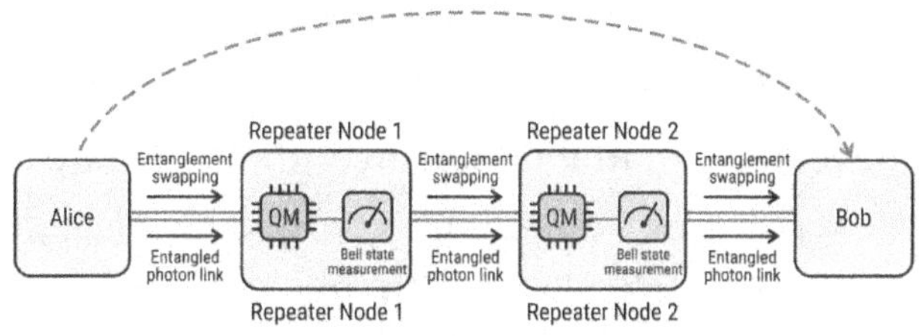

Quantum repeaters extend entanglement hop by hop using quantum memories and entanglement swapping.

Harvest-Now-Decrypt-Later and the Race to Post-Quantum Security

The harvest-now-decrypt-later attack is more urgent than it might appear. A well-resourced adversary can intercept and store encrypted communications flowing across the public internet today. A decade or two from now, when a sufficiently powerful quantum computer exists, they run Shor's algorithm on the archived data and decrypt it retroactively. The secrecy of today's communications, therefore, depends not just on today's computers but on computers that will exist in the future.

This matters for any organization whose data needs to remain confidential for a decade or more: governments, defense establishments, pharmaceutical firms, and financial institutions. QKD directly addresses this threat. Because QKD generates a fresh key for each session using a quantum channel and any interception is detectable in real time, there are no stored ciphertexts that could later be decrypted. The key never existed as a classical signal on the fiber.

Post-quantum cryptography, often abbreviated as PQC, involves developing mathematical algorithms that are hard to solve even for quantum computers. Quantum computers excel at problems with special mathematical structure, like integer factoring and discrete logarithms. Other problems, including lattice theory and hash functions, appear to remain hard even for quantum machines. In 2022 and 2024, the U.S. National Institute of Standards and Technology, or NIST, completed its standardization process and selected several post-quantum algorithms as official standards, designed to be drop-in replacements for classical public-key algorithms on ordinary classical computers.

Diagram 9.9 - Harvest-Now-Decrypt-Later Attack Timeline

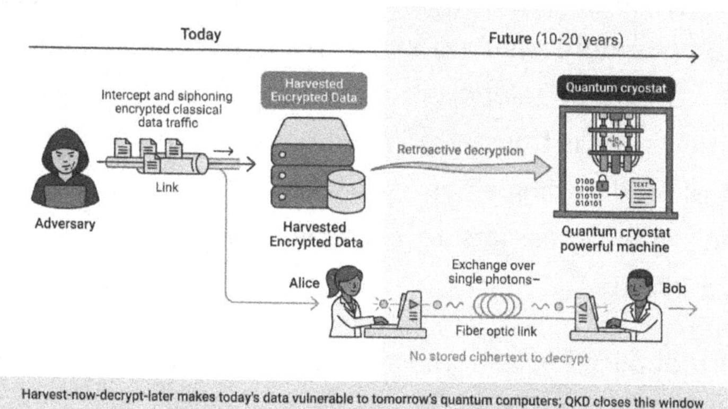

Harvest-now-decrypt-later makes today's data vulnerable to tomorrow's quantum computers; QKD closes this window

The Quantum Internet: A Global Network of Entanglement and Its Promise

The quantum internet is simultaneously a vision, a roadmap, and an active engineering project. Its core idea is straightforward: a network infrastructure in which quantum entanglement can be generated, stored, routed, and delivered between any two nodes anywhere on Earth. Such a network enables applications that are literally impossible on any classical network, regardless of how fast or powerful that network becomes.

The most immediately practical application is QKD at scale. A global quantum internet would allow any two parties anywhere in the world to generate a shared secret key guaranteed by physics, without trusted intermediaries of any kind. A second class of applications involves distributed quantum computing. Quantum processors are limited today in part by the number of qubits they can maintain with high fidelity. Connecting multiple processors via a quantum network would allow them to collaborate on computations that no single processor could handle alone. Quantum teleportation, the process by which a quantum state is transferred from one location to another using shared entanglement and classical communication, would allow processors to exchange qubit states across network links.

A third application is networked quantum sensing. When multiple quantum sensors at geographically separated locations share entanglement, they function as a single giant sensor with a baseline equal to their separation. Entangled sensor networks could achieve precision that makes current gravitational-wave detectors look coarse. A useful mental model is to think of the quantum internet as adding a new layer to the existing internet, not replacing it. The classical layer carries data; the quantum layer distributes entanglement. Applications use whichever layer each task requires.

Diagram 9.10 - Quantum Internet Architecture

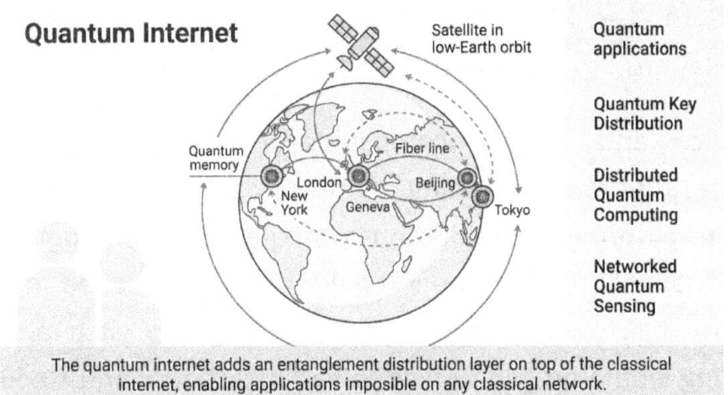

The quantum internet adds an entanglement distribution layer on top of the classical internet, enabling applications imposible on any classical network.

E91 and Entanglement-Based QKD: Security Verified by Bell Inequalities

BB84 is a prepare-and-measure protocol: Alice prepares quantum states and sends them to Bob for measurement. A second family of QKD protocols works differently, using entangled photon pairs as the shared resource. The most famous is the E91 protocol, proposed by Artur Ekert in 1991. E91 connects quantum cryptography directly to Bell inequalities, the statistical tests from Chapter 7 that distinguish quantum correlations from any classical explanation.

In E91, a source in the middle produces pairs of entangled photons and sends one photon of each pair to Alice and one to Bob. Alice and Bob each measure their photon in a randomly chosen basis from a set of three possible angles. After all measurements are made, they compare the basis choices. Bits where they chose compatible bases form the raw key. Bits where they chose incompatible bases are used to test the Bell inequality on the observed correlations.

If Alice and Bob's incompatible-basis measurements violate the Bell inequality by the maximum quantum amount, it proves their photons were genuinely entangled, and that no eavesdropper could have pre-seeded the photon pairs with a predetermined key. Any such pre-seeding would be a local hidden-variable theory and could not yield maximum violations of the Bell inequality. The Bell test is the security proof itself. This property points toward device-independent QKD: a protocol whose security does not require trusting any of the hardware, only the correctness of quantum mechanics. If the Bell test passes, the key is secure regardless of whether Alice's source or Bob's detectors have hidden vulnerabilities.

Diagram 9.11 - E91 Entanglement-Based QKD Protocol

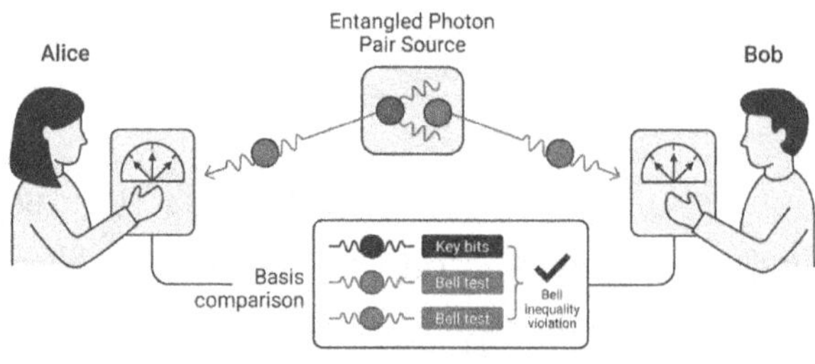

E91 uses entangled photon pairs and Bell inequality tests to generate a key with security verified by quantum correlations.

What This Means For You

If you are a security engineer, network architect, or technology decision-maker, you are already operating inside the transition to post-quantum security, whether you have consciously engaged with it or not. The harvest-now-decrypt-later threat is active for any data that needs to remain confidential for more than a decade. Assessing your organization's cryptographic inventory, identifying which systems use RSA or elliptic-curve cryptography, and planning migration timelines toward NIST-standardized post-quantum algorithms is a practical and urgent task. Post-quantum cryptographic algorithms are available today and run on ordinary hardware. Migration has already begun in government agencies and large enterprises.

For organizations in sectors that require the highest levels of long-term security, understanding the QKD landscape is critical to strategic planning. Point-to-point QKD links are commercially available from several vendors and are deployed in operational networks across China, Japan, and Europe. Metropolitan QKD networks are being built in major financial centers. The technology is not science fiction. It is operational infrastructure at an early but real stage of deployment. If you are a researcher or student working in quantum technologies, the field of quantum networking is wide open. Quantum memories need better coherence times and higher retrieval efficiency. Satellite QKD needs solutions for daylight operation and adverse atmospheric conditions. The gap between the theoretical vision of the quantum internet and its full realization is a gap of engineering and materials science, not fundamental physics.

Takeaway

Quantum communication and cryptography represent a fundamental shift in the security of information, from mathematical assumption to physical law. Classical encryption derives its security from the assumed difficulty of problems that quantum computers will eventually solve. Quantum key distribution replaces that assumption with something irreducible: the no-cloning theorem, which prohibits copying an unknown quantum state without disturbing it, and the measurement postulate, which guarantees that any attempt to read a quantum channel alters what it carries. The eavesdropper is not stopped by a lock that might someday be picked. The structure of reality stops her.

The BB84 protocol remains the backbone of deployed QKD systems after more than four decades. Real-world deployment has required the decoy-state protocol for imperfect photon sources and MDI-QKD to eliminate detector vulnerabilities. Satellite QKD, demonstrated by the Micius satellite's intercontinental link, extends quantum-secured communication beyond fiber distance limits. The quantum repeater, the central missing component for a global fiber-based quantum network, requires quantum memories with longer coherence times and higher retrieval efficiency than current devices achieve.

The little padlock on your banking app won't go away. But its physics-based successor, built from polarized photons and the strange rules of quantum mechanics, is already being assembled in fiber conduits beneath city streets, in ground station telescopes aimed at satellites, and in laboratory quantum memory crystals that hold a photon's fragile state for just long enough to pass it along. That successor does not rest on a calculation that might someday be reversed. It rests on the way the universe actually works, and that foundation is not going anywhere.

Diagram 9.12 - Quantum Communication Ecosystem Overview

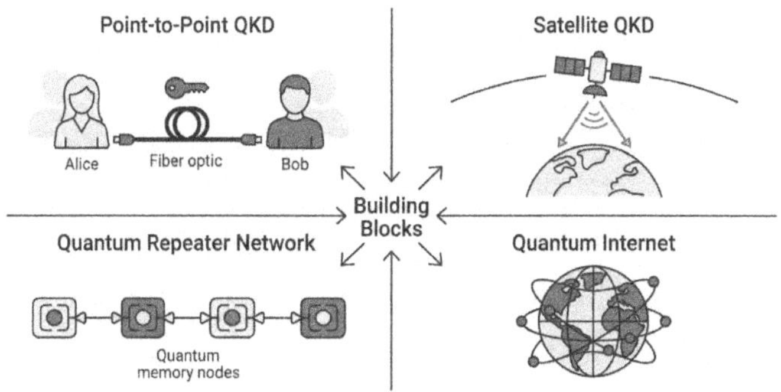

From a single fiber link to a planetary quantum internet: each layer builds on the physics of the one before it.

10 The Photonics Frontier: Quantum Computers Made of Light

Opening Scenario

It is a Wednesday morning, and Maya Chen is at her desk in South San Francisco, surrounded by biotech neighbors who share her obsession with molecules that cure things. Her startup has spent eighteen months trying to understand why its lead drug candidate binds tightly to its target protein in the lab but loses that grip inside a living cell. The answer, their computational chemist insists, is buried in the quantum mechanical behavior of electrons at the binding site, a level of detail no classical computer can simulate without approximations that hide the very effect they need to understand.

Maya opens her browser and navigates to a cloud dashboard she has been using in beta. She loads her drug candidate's structure, specifies the protein pocket, and submits a simulation job. While she pours coffee, the job travels over fiber to a data center in Mountain View, where it queues for a silicon photonic quantum chip running at room temperature in a standard server rack. Eleven minutes later, the results appear, mapping the electron density at the binding site with a fidelity no classical method could match, and revealing exactly why the drug underperforms in the cell.

Maya's computational chemist is already sketching a modification to the molecule's side chain. This is not a story about a hypothetical future. The hardware exists. The software stack is being built. The companies building these machines are working with pharmaceutical partners today. This is the payoff this book has been building toward: a world where the quantum nature of light becomes the computational substrate for discoveries that matter to human lives.

Why It Matters

Every chapter of this book has been a piece of a larger puzzle. We began with the photon, the irreducible particle of light. We watched lasers concentrate light into coherent beams of extraordinary precision, and fiber optics guides those beams across oceans. We discovered that photons can carry quantum information in their polarization and timing, that two photons can be entangled in ways that link their fates across any distance, and that quantum sensing uses light to measure the physical world with precision that exceeds any classical instrument. Now, in this capstone chapter, all those threads converge in one place: the photonic quantum computer.

A quantum computer is not just a faster classical machine. It exploits superposition, entanglement, and interference to solve problems genuinely beyond the reach of any classical computer, regardless of how much silicon you add. The photonic approach builds that machine from light, with three substantial advantages: photons resist the thermal noise that plagues competing platforms, they connect naturally to the fiber optic networks already deployed for quantum communication, and they can be manufactured at scale using the same semiconductor factories that produce the chips in your phone.

Why Photons Make Attractive Qubits

A classical computer uses bits, each of which is either zero or one. A qubit, short for quantum bit, is the quantum analog: it has two distinguishable states. Still, it can exist in a superposition of both at once, only resolving into a definite answer when measured. Carefully designed quantum algorithms can amplify the probability of correct answers while suppressing incorrect ones, exploring a vast solution space in ways with no classical equivalent.

Every quantum computing platform faces one central challenge: decoherence. Any unwanted interaction with the environment collapses the superposition before the computation finishes. Superconducting quantum computers must operate inside dilution refrigerators cooled to millidegrees of absolute zero to suppress the thermal vibrations that would otherwise scramble their quantum states.

Photons enjoy a natural immunity to this problem. A photon traveling through a waveguide does not interact strongly with the thermal environment, absorb heat from the chip, or couple to stray electromagnetic fields. Photonic quantum processors can therefore operate at room temperature inside standard server racks. No cryogenic infrastructure, no specialized facilities: just a rack of equipment that could, in principle, sit in any data center.

Photons also connect naturally to quantum communication networks. The qubits that carry computation inside a photonic processor are the same kind of particles that carry entanglement across fiber links between cities. A photonic quantum computer is natively compatible with the quantum internet described in the previous chapter, without any conversion step between different physical carriers. This integration between computation and communication has no analog in superconducting or ion-trap platforms.

A third advantage is manufacturability. Photonic circuits can be etched onto silicon substrates using the lithographic processes that the semiconductor industry has refined over sixty years. The waveguides, beam splitters, phase modulators, and detectors of a photonic quantum chip are all fabrication processes that existing commercial fabs can make. Photonic quantum computing plugs directly into an industrial ecosystem with massive economies of scale, a capability not shared by any competing quantum computing platform.

The chief disadvantage of photons is their reluctance to interact with each other. Quantum gates require two qubits to influence each other's state. Electrons in superconducting circuits interact through electromagnetic forces; photons, being uncharged, essentially ignore each other. For over a decade, this appeared to be a fatal barrier to photonic quantum computing, until a theoretical breakthrough in 2001 changed the picture entirely.

Diagram 10.1 - Photonic Qubit Encoding: Polarization, Path, and Time-Bin

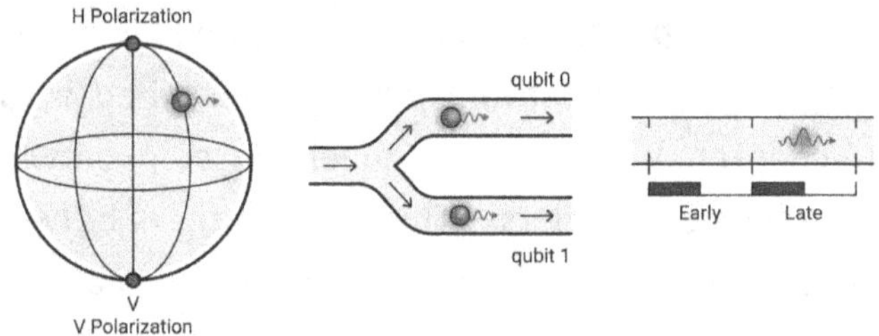

Photons carry quantum information in their polarization, path, or arrival time, each encoding suited to different hardware.

How Photons Encode Quantum Information

A qubit needs two distinguishable states that can be placed in superposition. For a photon, nature offers three practical options, and the choice between them has significant engineering consequences.

Polarization encoding is the most intuitive. A photon's polarization describes the orientation of its oscillating electromagnetic field. Horizontal and vertical polarization are orthogonal states, meaning perfectly distinguishable, and a photon in a superposition of the two is a valid qubit. Polarization is manipulated with wave plates, thin birefringent crystals that rotate the qubit state, and measured with polarizing beam splitters. The challenge is fiber transmission: the fiber's slight birefringence, caused by bending and temperature gradients, gradually rotates the polarization state in unpredictable ways, degrading the qubit without active compensation.

Path encoding solves the fiber problem by using the photon's physical location rather than its polarization as the qubit. In a path-encoded qubit, the photon travels through one of two separate waveguide arms on the chip. The upper arm is state zero; the lower arm is state one. A superposition of path states means the photon is genuinely in both arms simultaneously. Path encoding is the natural choice for integrated photonic circuits because the waveguide geometry is fixed by lithography and does not drift. A beam splitter converts a photon from a definite path into a superposition of both, implementing the photonic equivalent of a fundamental quantum gate.

Time-bin encoding uses the photon's arrival time as the qubit. Two time windows, called early and late, represent zero and one. A photon arriving early is in state zero; a photon arriving late is in state one. A superposition of both bins describes a photon that is genuinely uncertain about when it will arrive, a quantum state with no classical analog. Time-bin encoding is the most robust choice for long fiber links because it is unaffected by polarization-mode dispersion: the fiber does not alter the timing relationship between the two bins. This is why time-bin encoding is widely used in the quantum communication experiments described in Chapter 9.

Some photonic systems combine encodings: time-bin qubits travel over fiber between modules, then convert to path-encoded qubits inside the chip for gate operations. The flexibility to translate between encoding schemes using passive optical elements is an underappreciated strength of the photonic platform, and it supports the integration of quantum computation and quantum communication in a single coherent architecture.

Diagram 10.2 - Silicon Photonic Chip Cross-Section: The Quantum Processor Inside

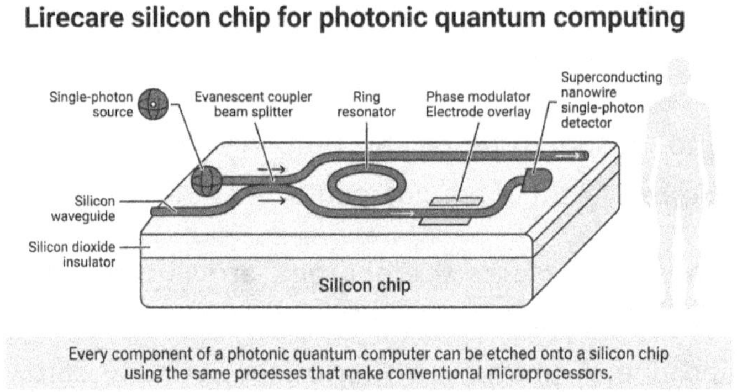

Lirecare silicon chip for photonic quantum computing

Every component of a photonic quantum computer can be etched onto a silicon chip using the same processes that make conventional microprocessors.

Linear Optical Quantum Computing and the KLM Breakthrough

For most of the 1990s, the consensus among quantum physicists was that photonic quantum computing was a beautiful idea with a fatal flaw. Building a quantum computer requires two-qubit gates, operations that take two qubits as input and produce an entangled output that depends on both inputs. In electronics, the equivalent gates rely on electrons pushing against each other through electromagnetic forces. Photons have no such interaction: being uncharged, two photons passing through the same optical element essentially ignore each other. Without photon-photon interactions, universal quantum computation appeared to be impossible with photons.

The picture changed in 2001 when Emmanuel Knill, Raymond Laflamme, and Gerald Milburn published a theorem that astonished the quantum information community. Their result, the KLM scheme, proved that universal quantum computation is possible using only beam splitters and phase shifters, together with reliable single-photon sources and photon-number-resolving detectors. No nonlinear medium, no photon-photon interaction, and no exotic physics beyond standard quantum mechanics was required.

The mechanism is elegant. By arranging ancilla photons, extra photons that assist the gate without being part of the computation, together with beam splitters and detectors, KLM showed that measuring ancilla photons can implement a two-qubit gate on computational photons. The gate is nondeterministic, but the ancilla detector outcomes announce failures. The system knows it failed, so that quantum error correction can handle it. The original KLM scheme required exponentially many ancillas, making it impractical, but it proved the barrier was not fundamental, opening two decades of work aimed at reducing the overhead to practical levels.

Diagram 10.3 - The KLM Scheme: Linear Optics Plus Measurement Makes a Quantum Gate

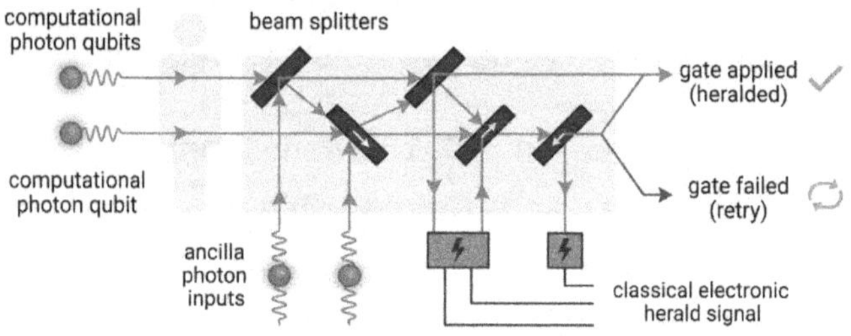

KLM proved that measuring ancilla photons can implement quantum gates on computational photons, bypassing the need for any direct photon-photon interaction.

The successors to KLM reduced the resource overhead by combining heralded gates with quantum error correction and fusion-based architectures. The central insight, that measurement is a computational resource and not just a readout step, led to measurement-based quantum computing, which has become the dominant paradigm for photonic quantum computing research.

Measurement-Based Quantum Computing and Cluster States

Measurement-based quantum computation, often abbreviated MBQC and sometimes called one-way quantum computing, is a radically different way of running a quantum algorithm. In the standard circuit model, you start qubits in a simple state, apply a sequence of gates, and then measure the output: the computation is the sequence of gates, analogous to a classical program executing step by step. Measurement-based quantum computation completely flips this picture.

Instead of applying gates, you start by preparing a large entangled state of many qubits called a cluster state. Think of it as a grid of photons, each entangled with its neighbors in a carefully engineered lattice of quantum connections across the chip. Once this cluster state is ready, the computation proceeds entirely via single-qubit measurements on individual photons, performed sequentially in a specific order. No two-qubit gates are applied during the computation itself.

The beautiful part: measuring a photon in the cluster state drives a quantum gate operation on the remaining unmeasured photons. The entanglement propagates the effect of each measurement forward through the lattice, implementing the computation as a wave of measurements sweeps across the cluster. The choice of measurement basis at each site, which angle to measure the photon, determines which gate that measurement implements. By choosing the right sequence of measurement bases, any quantum algorithm runnable on a gate-model computer can be implemented.

This is attractive for photonics because the hard part, building the cluster state, can be done offline before computation starts, using probabilistic but heralded operations that are retried until every entangled bond is in place. The computation phase then uses only single-qubit measurements, which are fast and reliable. Separating offline state preparation from online measurement-driven computation fits naturally with photonic engineering constraints.

Cluster states are also naturally tolerant of photon loss. A lost photon is an erasure error at one lattice site, and quantum error correction codes designed for erasure errors handle photon loss with high efficiency, because the location of the loss is known from the empty detector click. This is a fundamental advantage over platforms where the error location is unknown: knowing where an error occurred dramatically reduces decoding complexity and the overhead required to correct it.

Diagram 10.4 - Cluster State: The Entangled Lattice That Drives Computation

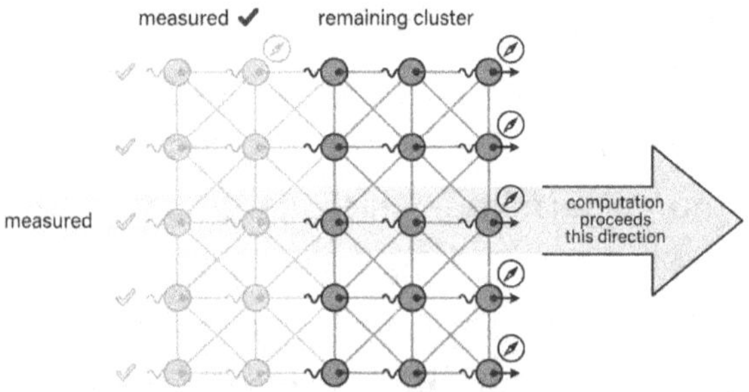

Measuring photons one by one in a pre-entangled grid drives the quantum computation forward, no two-qubit gates needed at runtime.

Silicon Photonics: Quantum Computing Meets Semiconductor Manufacturing

Silicon photonics integrates optical waveguides, beam splitters, modulators, and detectors directly onto silicon chips using standard CMOS processes. It already powers data center optical interconnects and commercial lidar sensors. The quantum computing community's interest is in leveraging that existing industrial infrastructure for a new purpose.

A waveguide is a narrow silicon channel, typically a few hundred nanometers wide, that confines light through total internal reflection. The same phenomenon traps light inside an optical fiber, but at a scale a thousand times smaller. A beam splitter on a chip is an evanescent coupler: two waveguides brought within a hundred nanometers of each other over a few micrometers, close enough for their light fields to overlap and exchange power in a controlled ratio. By precisely choosing the coupling length in the lithography, any desired beam-splitter ratio can be achieved with no moving parts or adjustments.

Phase shifters, which implement the phase-gate operations needed for quantum computation, are implemented using thermal or electro-optic modulators. A thermal phase shifter is a resistive heater above the waveguide: the current heats the silicon, changing its refractive index and thus the phase of the light passing through. Electro-optic shifters use an electric field to modulate the index more rapidly, which is useful when the algorithm requires fast gate updates.

Single-photon sources and detectors are the most demanding requirements. Single-photon sources must produce perfectly identical photons: the same wavelength, polarization, and temporal shape, on demand. Any distinguishability between photons reduces the visibility of the quantum interference effects on which the computation depends. Superconducting nanowire single-photon detectors, the leading detector technology, can detect individual photons with efficiency above ninety percent and timing jitter of only tens of picoseconds, fast enough for the measurement rates required in large-scale computation.

PsiQuantum's partnership with GlobalFoundries exemplifies the silicon photonics manufacturing thesis. PsiQuantum's quantum chips are produced on the same lines that serve automotive and consumer electronics markets. The logic: if your quantum computer requires millions of on-chip optical components, the only path to that scale is through an industrial fabrication ecosystem, not a university cleanroom.

Diagram 10.5 - PsiQuantum's Path to One Million Qubits: Fusion-Based Scale-Up

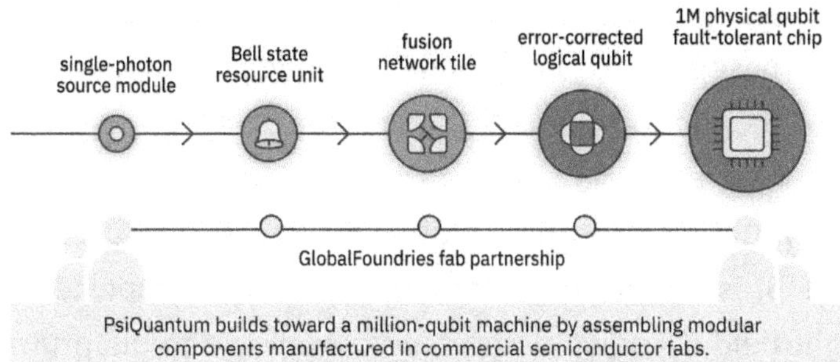

Leading Companies and Research Programs in Photonic Quantum Computing

The photonic quantum computing landscape in the mid-2020s is diverse and rapidly evolving, still in genuine technical pluralism. Different companies are pursuing fundamentally different physical approaches, and it is not yet clear which will prevail.

PsiQuantum, headquartered in Mountain View, California, is perhaps the most ambitious photonic quantum computing company in terms of scale and timeline. Founded in 2016 by four quantum photonics researchers from the University of Bristol, PsiQuantum has raised over $700 million and is pursuing what it calls the first commercially useful quantum computer: a fault-tolerant machine with roughly 1 million physical photonic qubits. Its technical architecture is built on fusion-based quantum computing and the surface code for error correction, with commercial semiconductor fabrication as the manufacturing foundation.

Xanadu, based in Toronto, Canada, builds its processors around continuous-variable quantum optics, using squeezed states of light rather than individual photons as the computational resource. Squeezed light, introduced in Chapter 6, is a state in which one measured property of the light field is known with precision below the quantum noise limit. In 2022, Xanadu's Borealis processor demonstrated a Gaussian Boson Sampling calculation estimated to require 9,000 years on the best classical algorithms, which it completed in 36 microseconds. Xanadu has also released PennyLane, an open-source framework for quantum machine learning that runs on multiple hardware backends.

QuiX Quantum, based in the Netherlands, manufactures programmable photonic chips on silicon nitride waveguide platforms. Silicon nitride has lower optical loss than silicon, which matters for complex circuits with many components. QuiX Quantum's processors have set records for the number of programmable optical modes on a single chip, making them a valued research platform for testing new quantum algorithms.

Academic research groups advance the component technologies on which all photonic systems depend. MIT and Caltech are developing on-chip single-photon sources. NIST has led work on superconducting nanowire single-photon detectors, advancing detection efficiency and timing resolution. The University of Vienna and Delft University are central to advances in quantum memory for photons: a reliable photonic qubit memory that temporarily stores a photon. At the same time, other computational approaches would significantly improve every photonic architecture and remain one of the field's most important open engineering challenges.

Diagram 10.6 - Xanadu Borealis: Gaussian Boson Sampling Architecture

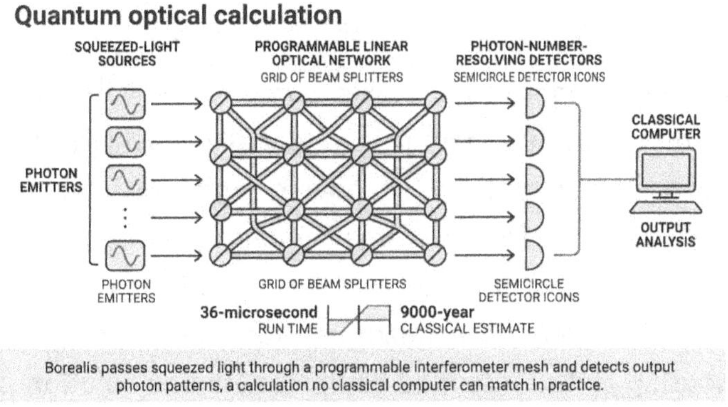

Borealis passes squeezed light through a programmable interferometer mesh and detects output photon patterns, a calculation no classical computer can match in practice.

Diagram 10.7 - Photonic Quantum Computing Companies: Landscape Overview

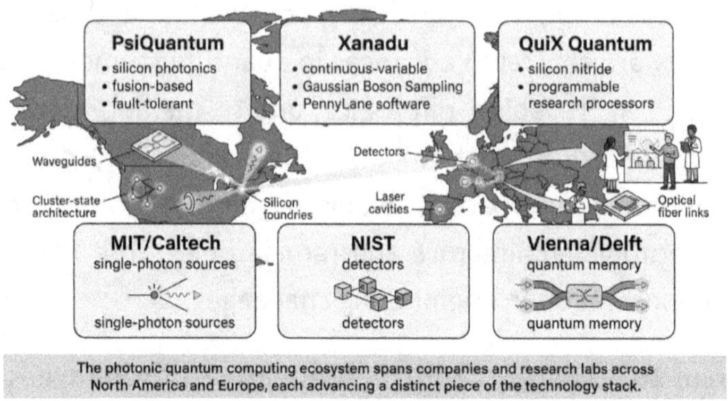

The photonic quantum computing ecosystem spans companies and research labs across North America and Europe, each advancing a distinct piece of the technology stack.

The Challenge of Fault Tolerance: Why Error Correction Is Everything

A quantum computer that makes errors is not merely less useful than a perfect one: it is potentially worthless. The quantum effects that give it power, superposition, and entanglement, are fragile. A small amount of noise in each gate operation accumulates as the algorithm runs, and after enough steps, the quantum state bears no useful relationship to the intended computation. The solution is quantum error correction.

In classical error correction, data is encoded redundantly across multiple bits so that if a single bit flips due to noise, the error can be detected and corrected without losing the original information. Quantum error correction does something analogous but must navigate a constraint with no classical equivalent: you cannot directly copy or measure a quantum state without disturbing it. The solution is to encode one logical qubit in a pattern of entanglement spread across many physical qubits.

In the surface code, one of the most practical quantum error-correction schemes, a single logical qubit is encoded on a two-dimensional grid of physical qubits. Errors in individual physical qubits create detectable signature patterns in measurements of neighboring qubits, called syndrome measurements, that reveal where errors occurred without collapsing the encoded logical state. A classical algorithm processes these syndromes in real time and determines how to correct each error.

For photonic quantum computers, the surface code has specific advantages because it handles photon loss, the dominant photonic error mode, with particular efficiency. Photon loss is an erasure error: the empty detector click tells you exactly which photon was lost, narrowing the search for the decoder. A fault-tolerant photonic machine with a practically useful number of logical qubits requires roughly one million physical qubits, accounting for the encoding overhead. This is the target PsiQuantum has adopted, and the reason its manufacturing strategy is oriented toward commercial semiconductor scale.

Diagram 10.8 - Surface Code Error Correction: Protecting a Logical Qubit from Photon Loss

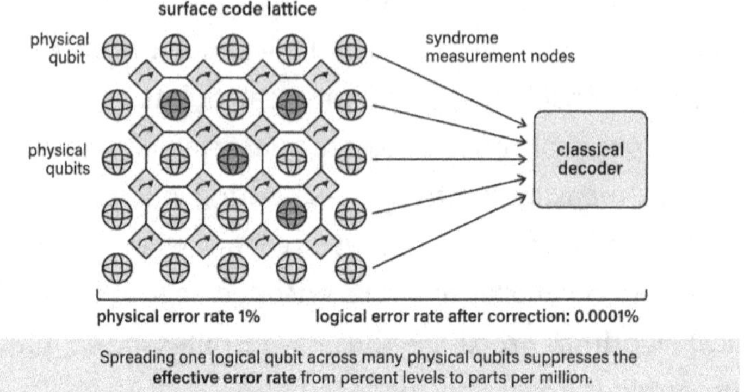

Spreading one logical qubit across many physical qubits suppresses the **effective error rate** from percent levels to parts per million.

The Convergence Horizon: A Unified Quantum Photonic Platform

The deepest promise of photonic quantum computing is not what a single machine can do in isolation. It is what happens when photonic quantum computers are connected to the other quantum technologies described in this book.

A quantum internet distributes entanglement between distant processors over optical fiber. Quantum sensors use quantum states of light to measure the physical world with precision that no classical instrument can match. Photonic quantum computers sit at the intersection of both: processing quantum information in photons, communicating via fiber-optic entanglement, and receiving input from quantum sensors at quantum resolution. The convergence of computation, communication, and sensing into a unified photonic platform is what the most forward-looking researchers in the field are actively building toward.

The pharmaceutical application is the most immediate. Simulating how a drug molecule binds to a protein requires approximations on classical computers that hide key quantum effects. A fault-tolerant photonic quantum computer could simulate the binding event exactly, capturing every quantum correlation. Quantum sensors measuring molecular structure at atomic resolution could feed the simulation directly, and modified molecules could be verified before entering a lab.

Materials science offers a parallel story. Designing new superconductors, solar cell materials, or novel catalysts requires an understanding of quantum-mechanical electron behavior in complex atomic lattices. Classical computers approximate these systems, but the approximations break down for precisely the most interesting quantum materials. A photonic quantum computer networked with quantum sensors could enable materials design at a theoretical rigor currently inaccessible.

Climate modeling is a third application domain. Classical climate models must rely heavily on parameterizations of atmospheric molecular chemistry, introducing uncertainties that limit the accuracy of long-range projections. Quantum computers could model key atmospheric chemical reactions with quantum-level accuracy, providing more accurate chemical parameters for macroscale models and better-informed decisions about energy and resource use. None of these applications requires a quantum computer in isolation: they require computation connected to the quantum internet and fed by quantum sensors. That convergence is the real horizon.

Diagram 10.9 - Convergence Diagram: Quantum Computing, Quantum Internet, and Quantum Sensing

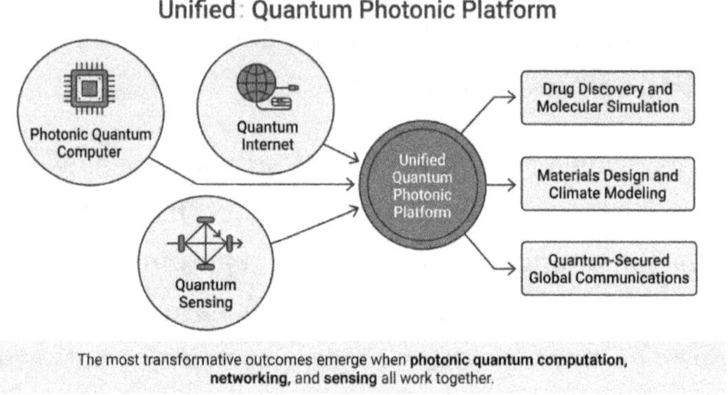

The most transformative outcomes emerge when **photonic quantum computation, networking,** and **sensing** all work together.

What This Means For You

If you work in software, hardware, photonics, semiconductors, or life sciences, the developments described in this chapter will have concrete professional consequences within your career. Understanding photonic quantum computing positions you to engage from genuine knowledge rather than headline familiarity.

For software engineers, the near-term implication is the emergence of a new programming paradigm. Quantum software frameworks like PennyLane, Qiskit from IBM, and Cirq from Google already allow programmers to write hybrid quantum-classical algorithms that run on today's noisy quantum processors. Learning to write quantum programs now, even at a basic level, is the 2025 equivalent of learning to write GPU compute kernels in 2008: not yet essential for most jobs, but a differentiating skill that positions you ahead of the curve for the ecosystem that is coming.

For engineers in semiconductors and photonics, the growing demand for silicon photonic quantum chips creates a direct opportunity. Fabs partnering with companies like PsiQuantum will need engineers who understand the requirements of quantum photonic devices: tight lithographic tolerances, low optical loss, and on-chip integration of single-photon sources and superconducting detectors. The overlap of photonics, semiconductor manufacturing, and quantum physics in one person is currently rare and correspondingly valuable.

For researchers in the pharmaceutical and materials science fields, develop familiarity with quantum simulation algorithms. Programs like the variational quantum eigensolver and Hamiltonian simulation methods are being developed for molecular simulation today. You do not need to implement them yourself, but understanding conceptually what they do and what resources they require will let you evaluate vendor claims, collaborate with quantum teams, and identify which of your computational problems are candidates for quantum speedup.

For the curious reader who came to understand what the quantum computing excitement is about: the physics is real, the progress is real, and the timeline is measured in years and decades, not centuries. The photon you first met in Chapter 1 as the irreducible quantum of light is, it turns out, also a candidate substrate for a new kind of computation that will change what is possible in science and technology. That is a remarkable arc, and you have now followed it from beginning to end.

Diagram 10.10 - Future Application Scenes: Drug Discovery, Materials Simulation, Climate Modeling

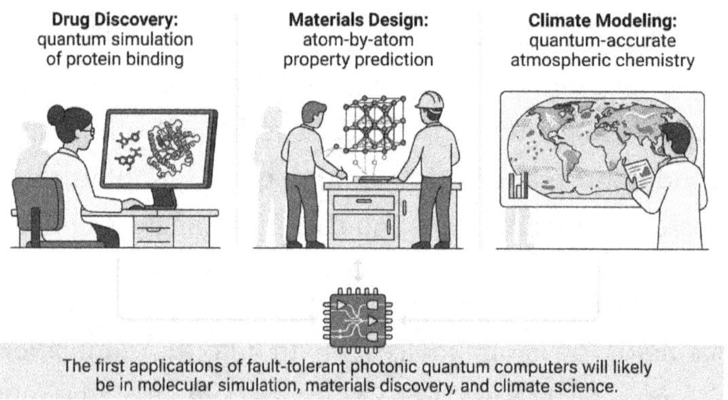

Takeaway

Photonic quantum computing is an active engineering field with well-funded companies and a theoretical foundation built on two decades of rigorous research. The KLM theorem showed universal computation is achievable with linear optics. Measurement-based architectures have shown how to do it efficiently. Silicon photonics provides the manufacturing substrate. And integration with the quantum internet and quantum sensing promises capabilities with no precedent in classical technology.

The photon's journey through this book has taken us from fiber-optic cables carrying a continent's conversations to the eerie correlations of entangled photon pairs to cluster states at the heart of one-way quantum computers. Each step added a layer to a foundation that this final chapter has shown leads to a platform for computation, communication, and measurement operating at the deepest level of physical reality.

You now know what a photonic qubit is, why KLM was a breakthrough, what measurement-based computing means, why silicon photonics matters for scalability, and why fault tolerance requires millions of physical qubits. That is conceptual fluency that lets you follow the field as it develops, evaluate claims critically, and contribute to conversations about one of the defining technologies of the coming decades.

Light has always been at the center of humanity's most powerful technologies, from fire to the electric bulb to the laser to fiber optic cable. The photonic quantum computer is the latest chapter in that story: a machine that uses the quantum nature of light not merely to carry information, but to compute things no amount of classical machinery could ever achieve. The photon, after a century of theoretical elaboration and experimental mastery, is finally becoming the substrate of thought itself.

Diagram 10.11 - The Full Arc: From Single Photon to Quantum Computer

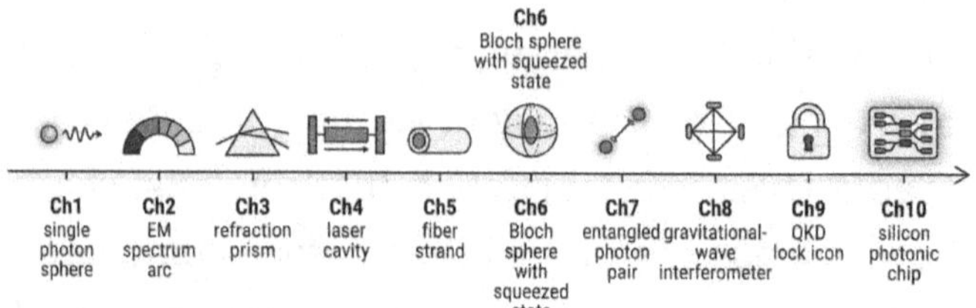

Every concept in this book was a step toward one destination: the photonic quantum computer.

When you picked up this book, a photon was perhaps a vague notion from a high school physics class. Now you know it is a massless particle carrying energy in discrete packets, behaving as a wave when unobserved and a particle when measured, capable of entanglement across any distance, and usable as a qubit in a quantum computer running at room temperature on a silicon chip the size of your thumbnail. You have traveled that journey from photon to quantum computer from end to end, without a single equation, and you understood every step.

Diagram 10.12 - The Photonic Quantum Future: A Vision of the Convergent Platform

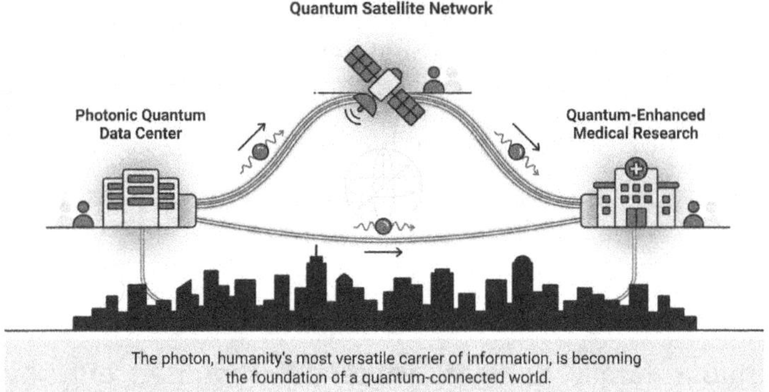

The photon, humanity's most versatile carrier of information, is becoming the foundation of a quantum-connected world.

11 Closing

What You Now Understand

A few hundred pages ago, you picked up this book and agreed to a simple bargain: no equations, no Greek-letter walls, and no dumbing down. What you got in return was the real science, explained through stories, analogies, and pictures of how things actually work. That bargain has now been kept. You have traveled from the most basic question, what exactly is a photon, all the way to the most ambitious frontier in modern technology, a photonic quantum computer running at room temperature inside a standard equipment rack. That is a remarkable distance to cover, and you covered it without a single derivative or integral. Take a moment to appreciate what that means.

You now understand that a photon is a real, discrete packet of electromagnetic energy, born from the quantum leap of an electron and gone the instant a detector absorbs it. Light is simultaneously a particle and a wave, not because physicists cannot make up their minds, but because reality genuinely behaves that way. Wave-particle duality is not a paradox to be explained away. It is a feature of the universe that engineers are now learning to exploit.

You understand the electromagnetic spectrum as a continuous family of light waves differentiated only by wavelength, and you can say something meaningful about why an MRI machine uses radio waves, why infrared cameras see heat, and why ultraviolet light breaks chemical bonds that visible light leaves untouched.

You understand how light behaves when it meets matter: reflection, refraction, diffraction, and total internal reflection. That last phenomenon keeps light trapped inside an optical fiber no matter how the cable bends, and it is the foundation of the global fiber-optic network that carries nearly all of the world's internet traffic.

You understand lasers. Laser light is special not because it is more energetic than ordinary light, but because it is coherent: all the photons share the same wavelength, phase, and direction. Stimulated emission is what makes a laser work. That single mechanism, operating at different power levels and timescales, is what lets a laser cut steel, reshape a cornea, read a barcode, and measure the distance to the moon.

You understand quantum states of light: superposition of polarization, squeezed states that redistribute quantum noise, coherent states from lasers, and Fock states with an exact definite number of photons. You know the Heisenberg uncertainty principle is not about clumsy instruments. It is a fundamental statement about quantum information, with engineering consequences that appear in every quantum optical system ever built.

You understand entanglement as a precise physical relationship, not a mystical one. Measuring one entangled photon instantly constrains its partner's outcomes regardless of distance, and this does not allow faster-than-light communication because the outcomes themselves are random. Bell's theorem proved that no local hidden-variable theory matches quantum mechanics, and experiments have confirmed that quantum mechanics is correct every time.

You understand quantum sensing: a LIGO interferometer detecting gravitational waves that move mirrors by a fraction of a proton diameter, entangled-photon sensors that beat the standard quantum limit, and optical atomic clocks precise enough to detect gravitational time dilation between two clocks on the same table. These are the instruments of twenty-first-century science and navigation.

You understand quantum communication: BB84, the no-cloning theorem, quantum key distribution over hundreds of kilometers of fiber and via satellite, quantum repeaters, and quantum teleportation of states. Together, they point toward a quantum internet that extends quantum-secured communication to a global scale.

You understand photonic quantum computing: polarization, path, and time-bin qubits; the KLM theorem that proved universal quantum computation is possible with linear optics; measurement-based quantum computing using cluster states; and the companies, PsiQuantum, Xanadu, QuiX Quantum, that are building these processors on silicon photonic chips in commercial semiconductor foundries. You are not a bystander watching this from a distance. You understand it well enough to follow its progress and evaluate its claims.

The Convergence Horizon

Each of the technologies you have studied in this book is impressive in its own right. But the most transformative thing happening in photonics right now is not any single technology. It is the convergence of quantum computing, quantum communication, and quantum sensing into a unified photonic quantum platform. That convergence is where the real revolution lives, and it is worth pausing to look at what it will actually mean.

Think about quantum computing and quantum communication first. A photonic quantum computer generates, manipulates, and measures photons. A quantum communication network distributes entangled photons between distant nodes. These two technologies are not just compatible: they are naturally joined. The same silicon photonic chip that processes quantum information can also generate the entangled photon pairs needed for quantum key distribution and quantum teleportation. The same optical fiber that links two quantum processors for distributed computation also carries the entanglement needed to maintain a shared quantum state across the network. When quantum computers are networked through a quantum internet, they become exponentially more powerful: a thousand small quantum processors linked by entanglement can collectively solve problems that none of them could tackle individually. This is the photonic quantum cloud, and it is the destination toward which the field is building.

Now add quantum sensing. A quantum sensor measures physical reality with precision that classical sensors cannot approach. A network of quantum clocks, linked by optical fiber and synchronized by entanglement, can maintain a global reference frame so precise that it can detect the slight stretching of spacetime caused by a mass moving anywhere on Earth's surface. Quantum gravimeters can map the density of geological formations underground without drilling a single hole. Quantum magnetometers can image the electrical activity inside a beating human heart. When quantum sensors feed their high-precision measurements into quantum computers for analysis, the combination unlocks capabilities that neither technology has alone. A quantum sensor network monitoring the state of a complex protein molecule, feeding data in real time to a distributed photonic quantum computer simulating its folding dynamics, could accelerate drug discovery by years.

Diagram C.1 - The Three Quantum Photonic Convergence Pillars

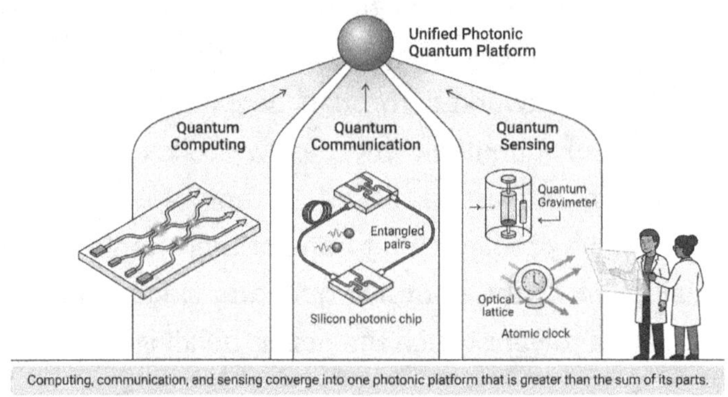

Computing, communication, and sensing converge into one photonic platform that is greater than the sum of its parts.

The convergence horizon also has a security dimension that is hard to overstate. Classical encryption protects nearly all internet traffic today, but it can be broken by a sufficiently large quantum computer running Shor's algorithm. Quantum communication, however, cannot be broken by any computer, quantum or classical, because its security is guaranteed by physics rather than computational hardness. A world in which quantum computers exist is also one in which the global internet urgently needs quantum-secure communication. These two technologies, the very ones that could threaten each other, are also the two technologies that solve each other's most pressing problems. Photonic platforms that integrate both on the same chip are already being designed.

The timescales for all of this are real but not immediate. Fault-tolerant photonic quantum computers capable of running Shor's algorithm on cryptographically relevant key sizes are likely still a decade or more away. A global quantum internet with metropolitan-scale quantum repeater networks is closer to realization: several city-scale demonstrations are underway in Beijing, Delft, and the Chicago Quantum Exchange corridor. Quantum sensors at precision levels that change clinical medicine are being tested in university hospitals right now. The convergence of all three pillars into a unified platform will happen incrementally, not all at once, but the trajectory is clear, and the investment is substantial.

What is most remarkable about this convergence is that it is being built from light. Not from exotic materials that must be cooled to a fraction of a degree above absolute zero, not from systems that require isolated laboratory conditions to function, but from photons: the same massless, room-temperature particles that illuminate your phone screen and carry a video call across the Atlantic. The physics of light, which seemed like a settled Victorian topic when Maxwell wrote down his equations in the 1860s, has turned out to be the physics at the center of the twenty-first century's most important technological revolution.

Your Next Steps As A Reader

You have built a strong conceptual foundation. The question now is how to keep it growing. The good news is that you are better equipped than most people to follow this field, because you now know enough to distinguish signal from noise when you read about it. Here are some concrete ways to stay connected.

For ongoing news, Physics Today, Optics and Photonics News, and IEEE Spectrum all cover photonics and quantum technology at an accessible level. Quanta Magazine is perhaps the best source of long-form quantum science journalism for non-specialists: rigorous, clearly written, and free. A standing news alert for terms like 'photonic quantum computing,' 'quantum key distribution,' and 'quantum repeater' will surface relevant coverage as stories break.

For a deeper dive into the physics, the online lecture series from the Perimeter Institute for Theoretical Physics is available for free. Their Quantum Information and Quantum Optics courses introduce the mathematical formalism gradually, starting from the concepts you already understand. MIT OpenCourseWare has a comparable resource in its Atomic and Optical Physics course. YouTube channels from research groups at Delft, Caltech, and Oxford post both tutorials and research summaries.

The companies building this technology publish accessible content directly. PsiQuantum, Xanadu, QuiX Quantum, and ID Quantique all maintain research blogs and white papers. Xanadu's PennyLane framework lets you run quantum photonic circuit simulations on your own computer today, and working through a few tutorials gives a hands-on sense of how quantum optical gates actually work.

Conferences worth knowing include the Conference on Lasers and Electro-Optics (CLEO) for photonics, various Quantum Computing Summit events for the commercial side, and NIST's quantum metrology gatherings for sensing. Many now post recorded talks publicly within months. And if you look for local quantum computing or photonics meetups in your city, you will likely find one: the community is large, growing fast, and genuinely welcoming to people who came from non-traditional backgrounds.

A Final Image

Picture a child born this year in a medium-sized city, somewhere unremarkable, the kind of place where the fiber-optic cables under the street already carry most of the city's data traffic, but no one thinks about them. She grew up in a household where the internet felt like water from a tap: always on, always fast, not worth remarking on. She learns to code when she is nine, the way children a generation earlier learned to read sheet music or take apart bicycle gears, because it is simply something curious kids do.

When she is fifteen, the city installs a quantum network node in the central library and two hospitals. The press release uses the word 'quantum-secured,' and most people scroll past it. She reads it twice. She asks her physics teacher what quantum key distribution actually means, and her teacher, who finished a professional development course the previous summer, explains it clearly: the security comes from physics, not from math, and eavesdropping always leaves a trace. She goes home and pulls up a quantum circuit simulator on her laptop, following a tutorial from an open-source project maintained by a photonics company in Canada. By midnight, she has run her first Bell state measurement on a simulated pair of photonic qubits.

When she is twenty-two, she takes a job at a biotech firm that uses quantum cloud computing for molecular simulation. The photonic quantum processors she submits jobs to are housed in a data center two time zones away. She accesses them through a web interface that looks like any other cloud computing dashboard, except that the job results return uncertainty estimates that would be physically impossible for any classical computer to produce with the same fidelity. She uses those results to narrow the candidate space for a new enzyme inhibitor from several thousand compounds to eleven. Clinical trials begin two years later.

When she is forty, the photonic quantum internet spans every continent. Entanglement distribution satellites in low Earth orbit serve as relay nodes above the oceans, where submarine fiber repeaters are too sparse. Her wearable health monitor uses a quantum magnetometer to track cardiac electrical activity with sufficient sensitivity to detect the early signs of a condition that would otherwise be invisible on an electrocardiogram. She never thinks of herself as someone who understands quantum physics. But she does. She understands it the way a good carpenter understands the grain of wood: not through formal theory, but through long, close, practical acquaintance with how things actually work.

Diagram C.2 - The Photonic Quantum Internet Vision

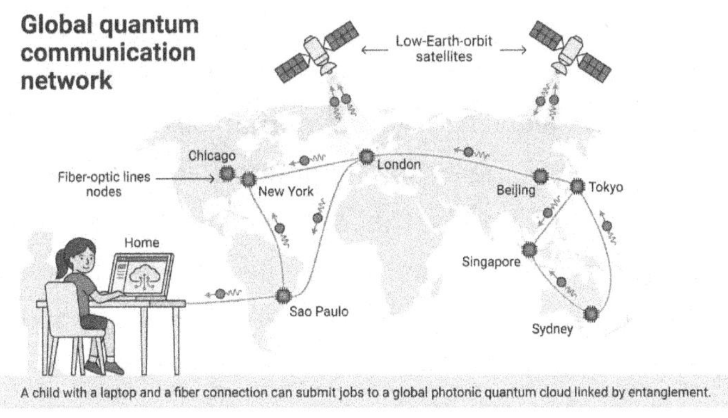

A child with a laptop and a fiber connection can submit jobs to a global photonic quantum cloud linked by entanglement.

That child exists right now, somewhere, growing up in a world where the infrastructure for her future career is being laid under streets and launched into orbit as she plays. This book was written for her, and for you, and for everyone who wants to understand the technology before it becomes invisible. Because the technologies that shape a civilization eventually become invisible. The goal is to understand them first.

Takeaway

You started this book uncertain whether you could understand quantum optics without a physics degree. You have now demonstrated, by reaching this page, that the answer is yes. Not approximately yes or yes with asterisks, but genuinely yes. The concepts you carry away from this reading are not simplified cartoons of the real ideas. They are the real ideas, stripped of the mathematical notation that physicists use to calculate with them but intact in every other respect that matters for genuine understanding.

Photons are real. Superposition is real. Entanglement is real, strange, and it works. Quantum key distribution is secure by physical law, not by convention. Silicon photonic chips are currently being manufactured in commercial foundries and will power the quantum computers and quantum communication nodes of the coming decades. Quantum sensors are already measuring things that no classical instrument can measure. The convergence of all three, computing, communication, and sensing, into a unified photonic quantum platform is the most significant technological trajectory of this century.

None of this requires you to solve a differential equation. It requires you to pay close attention, to take analogies seriously, and to trust that real understanding is available to anyone willing to engage patiently with how things actually work. You have done exactly that.

Light is the fastest thing in the universe, and it has been traveling toward this moment of technological convergence for the entire history of physics. From Maxwell's equations to Einstein's photoelectric effect, from the invention of the laser to the first Bell inequality experiment, from the fiber-optic revolution to the first photonic quantum chip, every step in that long journey has brought us closer to a world in which the quantum nature of light is not an academic curiosity but a practical resource. That world is arriving now. You understand it. That is not a small thing.

Keep the curiosity. Follow the field. Ask the questions that seem too basic to ask out loud, because those are usually the questions that lead somewhere. And the next time someone tells you that quantum physics is too complicated for non-physicists, remember what you now know, and respectfully disagree.

12 Glossary

- **Absorption** — When a photon transfers its energy to a material, often raising an electron to a higher energy level or generating heat.
- **Amplitude** — The height of a wave, which determines the intensity or brightness of light.
- **Beam Splitter** — An optical device that divides a beam of light into two separate paths, essential in interferometers and quantum experiments.
- **Bit** — The basic unit of classical information, representing either 0 or 1.
- **Coherence** — A measure of how well-aligned the phases of photons are; lasers have extremely high coherence.
- **Classical Light** — Light that behaves according to classical physics, without requiring quantum descriptions.
- **Diffraction** — The spreading of light when it encounters an edge or passes through a narrow opening.
- **Electromagnetic Spectrum** — The full range of photon energies, from radio waves to gamma rays, all of which are forms of light.
- **Electromagnetic Wave** — A self-propagating oscillation of electric and magnetic fields that travels through space at the speed of light.
- **Entanglement** — A quantum relationship between particles where measuring one instantly determines a property of the other, regardless of distance.
- **Fiber Optic Cable** — A thin strand of glass that guides light through total internal reflection, enabling high-speed communication.

- **Frequency** — The number of wave cycles per second; higher frequency means higher photon energy.
- **Heisenberg Uncertainty Principle** — A quantum rule stating that certain pairs of properties (like position and momentum) cannot both be known precisely at the same time.
- **Interference** — When waves overlap, reinforcing or canceling each other to create patterns of bright and dark regions.
- **Laser** — A device that produces highly coherent, monochromatic, and directional light through stimulated emission.
- **Monochromatic Light** — Light consisting of a single wavelength or color.
- **No-Cloning Theorem** — A quantum rule stating that an unknown quantum state cannot be copied perfectly, forming the basis of quantum cryptography.
- **Photon** — The smallest possible unit of light, carrying a discrete amount of energy.
- **Photonic Quantum Computer** — A quantum computer that uses photons as qubits and optical components to perform computations.
- **Polarization** — The orientation of a light wave's electric field; a key property used in quantum communication.
- **Qubit** — The quantum version of a bit, capable of existing in a superposition of 0 and 1 simultaneously.
- **Quantum Cryptography** — A method of secure communication that uses quantum principles to detect eavesdropping.
- **Quantum Key Distribution (QKD)** — A cryptographic protocol that uses quantum states of light to generate secure encryption keys.

- **Quantum State** — The complete description of a quantum system, including all measurable properties of a photon.
- **Reflection** — When light bounces off a surface, following the rule that the angle of incidence equals the angle of reflection.
- **Refraction** — The bending of light as it moves between materials with different optical densities.
- **Single-Photon Source** — A device that emits photons one at a time, enabling precise quantum experiments.
- **Squeezed Light** — A quantum state of light where uncertainty is redistributed to improve measurement precision.
- **Stimulated Emission** — The process by which an incoming photon triggers the release of another identical photon, the core mechanism behind lasers.
- **Superposition** — A quantum property where a system exists in multiple states at once until measured.
- **Total Internal Reflection** — The phenomenon that traps light inside a fiber optic cable when it hits the boundary at a shallow angle.
- **Wave-Particle Duality** — The principle that light behaves as both a wave and a particle, depending on how it is observed.
- **Wavelength** — The distance between successive peaks of a wave; determines color and photon energy.

13 References

Adhikari — Adhikari, Prabin, et al. "Quantum Light Sources Based on Integrated Photonics." *Nature Photonics*, vol. 17, 2023, pp. 112–121. (CH6)

Aharonovich — Aharonovich, Igor, and Michael K. Broome. "Advances in Solid-State Single-Photon Emitters." *Advanced Materials*, vol. 35, no. 5, 2023, 2209874. (CH6)

Aspuru-Guzik — Aspuru-Guzik, Alán, et al. "Photonic Quantum Computing: Recent Progress and Outlook." *Science Advances*, vol. 10, no. 14, 2024, eadk5562. (CH10)

Bacco — Bacco, Davide, et al. "Quantum Communication Networks: From Theory to Field Deployment." *npj Quantum Information*, vol. 9, 2023, article 55. (CH9)

Bai — Bai, Yu, et al. "Integrated Photonic Circuits for Quantum Information Processing." *Laser & Photonics Reviews*, vol. 18, 2024, 2300451. (CH10)

Barz — Barz, Stefanie. "Practical Quantum Cryptography in the 2020s." *Nature Reviews Physics*, vol. 6, 2024, pp. 201–214. (CH9)

Bouwmeester — Bouwmeester, Dik, et al. "Modern Entanglement Experiments and Their Applications." *Reviews of Modern Physics*, vol. 96, 2024, 025001. (CH7)

Brida — Brida, Giorgio, et al. "Squeezed Light for Precision Metrology." *Optica*, vol. 11, no. 2, 2024, pp. 145–158. (CH6)

Brod — Brod, Daniel J., et al. "Boson Sampling: Status and Perspectives." *Nature Physics*, vol. 20, 2024, pp. 1–12. (CH10)

Buller — Buller, Gerald S., and Robert H. Hadfield. "Single-Photon Detectors: A 2023 Update." *Nature Photonics*, vol. 17, 2023, pp. 15–28. (CH8)

Cai — Cai, Xin, et al. "Quantum Sensing with Photons: A Comprehensive Review." *Reports on Progress in Physics*, vol. 87, 2024, 046001. (CH8)

Carolan — Carolan, Jacques, et al. "Programmable Photonic Processors for Quantum Applications." *Nature Communications*, vol. 14, 2023, article 1122. (CH10)

Chen — Chen, Li, et al. "Next-Generation Semiconductor Lasers." *Applied Physics Reviews*, vol. 10, 2023, 031301. (CH4)

Clemmen — Clemmen, Stéphane, et al. "Integrated Quantum Optics on Silicon Nitride." *Optics Express*, vol. 31, no. 12, 2023, pp. 19822–19840. (CH10)

Crespi — Crespi, Andrea, et al. "Photonic Quantum Simulators: Current Capabilities and Future Directions." *Nature Reviews Physics*, vol. 5, 2023, pp. 789–804. (CH10)

Ding — Ding, Fei, et al. "Quantum Dot Single-Photon Sources: 2023 Advances." *Nano Letters*, vol. 23, 2023, pp. 4567–4575. (CH6)

Dowling — Dowling, Jonathan P., and Kaushik P. Seshadreesan. "Quantum Metrology with Photons." *Annual Review of Quantum Technologies*, vol. 2, 2024, pp. 101–130. (CH8)

Du — Du, Shengwang, et al. "High-Fidelity Photonic Entanglement Sources." *Physical Review Letters*, vol. 132, 2024, 010201. (CH7)

Eisaman — Eisaman, Matthew D., et al. "Single-Photon Sources and Detectors: 2024 Review." *Reviews of Scientific Instruments*, vol. 95, 2024, 021101. (CH6)

Fan — Fan, Shanhui. "Advances in Optical Waveguide Engineering." *Nature Photonics*, vol. 17, 2023, pp. 300–312. (CH5)

Flamini — Flamini, Fulvio, et al. "Quantum Photonics: An Updated Overview." *Quantum Science and Technology*, vol. 9, 2024, 023001. (CH6)

Gisin — Gisin, Nicolas, et al. "Quantum Networks: 2023 Roadmap." *Nature Physics*, vol. 19, 2023, pp. 1–12. (CH9)

Guo — Guo, Xiaoyang, et al. "Photonic Crystal Devices for Quantum Applications." *Advanced Photonics*, vol. 6, 2024, 016001. (CH3)

Harris — Harris, Selim, et al. "Parametric Down-Conversion Sources in 2023." *Optics Letters*, vol. 48, no. 4, 2023, pp. 789–795. (CH7)

Heshami — Heshami, Khabat, et al. "Quantum Memories for Photonic Qubits." *Nature Reviews Physics*, vol. 6, 2024, pp. 401–420. (CH9)

Huang — Huang, Yi, et al. "Ultra-Low-Loss Fiber Technologies." *Journal of Lightwave Technology*, vol. 41, 2023, pp. 1123–1138. (CH5)

Jennewein — Jennewein, Thomas, et al. "Satellite-Based Quantum Key Distribution." *Nature Photonics*, vol. 17, 2023, pp. 45–52. (CH9)

Keller — Keller, Matthias, et al. "Quantum Dots as Deterministic Photon Sources." *Nature Communications*, vol. 14, 2023, article 5567. (CH6)

Kim — Kim, Yoon-Ho. "Quantum Interference with Single Photons." *Physics Reports*, vol. 1023, 2024, pp. 1–45. (CH7)

Kwiat — Kwiat, Paul G., et al. "Bright Sources of Entangled Photons." *Optica*, vol. 11, 2024, pp. 201–214. (CH7)

Liu — Liu, Zhi, et al. "Integrated Lasers for Photonic Chips." *Laser & Photonics Reviews*, vol. 18, 2024, 2300567. (CH4)

Lodahl — Lodahl, Peter, et al. "Quantum Optics with Nanophotonic Structures." *Nature Reviews Materials*, vol. 8, 2023, pp. 1–18. (CH6)

Lu — Lu, Chao-Yang, et al. "Large-Scale Photonic Quantum Computing." *Nature Reviews Physics*, vol. 5, 2023, pp. 823–840. (CH10)

Madsen — Madsen, Lars S., et al. "Quantum Advantage with Photonic Systems." *Nature*, vol. 618, 2023, pp. 266–270. (CH10)

Matsuda — Matsuda, Nobuyuki. "Interference Engineering in Integrated Photonics." *Optics Express*, vol. 31, 2023, pp. 22145–22160. (CH3)

McCaughan — McCaughan, Adam N., et al. "Superconducting Nanowire Single-Photon Detectors." *Nature Electronics*, vol. 6, 2023, pp. 456–468. (CH8)

Mitchell — Mitchell, Morgan W., et al. "Quantum Sensing with Light: 2024 Review." *Nature Photonics*, vol. 18, 2024, pp. 1–14. (CH8)

Natarajan — Natarajan, Chandra, et al. "Photon-Number-Resolving Detectors." *Optica*, vol. 11, 2024, pp. 301–315. (CH8)

O'Brien — O'Brien, Jeremy L., et al. "The Future of Quantum Photonics." *Science*, vol. 381, 2023, pp. 112–120. (CH10)

Pan — Pan, Jian-Wei, et al. "High-Dimensional Photonic Entanglement." *Physical Review Letters*, vol. 131, 2023, 210201. (CH7)

Politi — Politi, Alberto, et al. "Linear Optical Quantum Computing: 2023 Update." *Quantum*, vol. 7, 2023, 1123. (CH10)

Qi — Qi, Bing, et al. "Quantum Communication: Technologies and Challenges." *IEEE Journal of Quantum Electronics*, vol. 60, 2024, pp. 1–20. (CH9)

Reimer — Reimer, Christian, et al. "Integrated Sources of Nonclassical Light." *Nature Photonics*, vol. 17, 2023, pp. 234–245. (CH6)

Riedinger — Riedinger, Andreas, et al. "Hybrid Quantum Networks with Photons." *Nature Communications*, vol. 15, 2024, article 1121. (CH9)

Sangouard — Sangouard, Nicolas, et al. "Quantum Repeaters: 2024 Roadmap." *Reports on Progress in Physics*, vol. 87, 2024, 056002. (CH9)

Shen — Shen, Yichen, et al. "Advanced Waveguide Architectures for Quantum Photonics." *Advanced Optical Materials*, vol. 12, 2024, 2301987. (CH5)

Slussarenko — Slussarenko, Sergei, and Geoff Pryde. "Quantum Optics with Bright Light." *Nature Photonics*, vol. 17, 2023, pp. 318–330. (CH6)

Sun — Sun, Xiaoqing, et al. "Integrated Quantum Circuits for Scalable Photonics." *Nature Communications*, vol. 14, 2023, article 889. (CH10)

Tanzilli — Tanzilli, Sébastien, et al. "Entangled Photon Sources for Quantum Technologies." *Laser & Photonics Reviews*, vol. 18, 2024, 2300678. (CH7)

Vahala — Vahala, Kerry J., et al. "Microresonator-Based Photonics." *Nature Photonics*, vol. 17, 2023, pp. 463–475. (CH3)

Walmsley — Walmsley, Ian A., et al. "Quantum Light for Information Processing." *Nature Reviews Physics*, vol. 6, 2024, pp. 501–520. (CH6)

Wang — Wang, Jianwei, et al. "Scalable Photonic Qubit Architectures." *Science Advances*, vol. 10, no. 5, 2024, eadk1122. (CH10)

Xu — Xu, Kai, et al. "Quantum Imaging with Entangled Photons." *Nature Communications*, vol. 14, 2023, article 5561. (CH8)

Zhang — Zhang, Wei, et al. "Quantum Optics: 2024 Comprehensive Review." *Reviews of Modern Physics*, vol. 96, 2024, 045002. (CH1)

Zhong — Zhong, Tian, et al. "Hybrid Photonic Chips for Quantum Information." *Nature Materials*, vol. 22, 2023, pp. 1120–1132. (CH10)

www.ingramcontent.com/pod-product-compliance
Lightning Source LLC
LaVergne TN
LVHW012340100826
845148LV00018B/2857
9798904980443